THE INVASION OF
HITLER'S
THIRD REICH

Colonel-General Alfred Jodl, Chief of Staff under the Doenitz Regime, signs the document of unconditional surrender.

Comet tanks of the 2nd Fife and Forfar Yeomanry, 11th Armoured Division, crossing the Weser at Petershagen, Germany, 7 April 1945.

THE INVASION OF HITLER'S THIRD REICH

PATRICK DELAFORCE

FONTHILL

*For my sweet wife with very many thanks
for her help with this book*

Fonthill Media Limited
Fonthill Media LLC
www.fonthillmedia.com
office@fonthillmedia.com

First published in 2011 under the title *Invasion of the Third Reich, War and Peace*.
This shortened edition published in the United Kingdom and the United States of America 2014

British Library Cataloguing in Publication Data:
A catalogue record for this book is available from the British Library

Copyright © Patrick Delaforce 2014

ISBN 978-1-78155-325-1

The right of Patrick Delaforce to be identified as the author of this work has been asserted by
him in accordance with the Copyright, Designs and Patents Act 1988.

Typeset in Sabon by Fonthill Media Limited
Printed and bound by CPI Group (UK) Ltd, Croydon, CR0 4YY

Contents

The last defenders – Hitler decorating Hitler Youth who won the Iron Cross in battle, 30 March 1945.

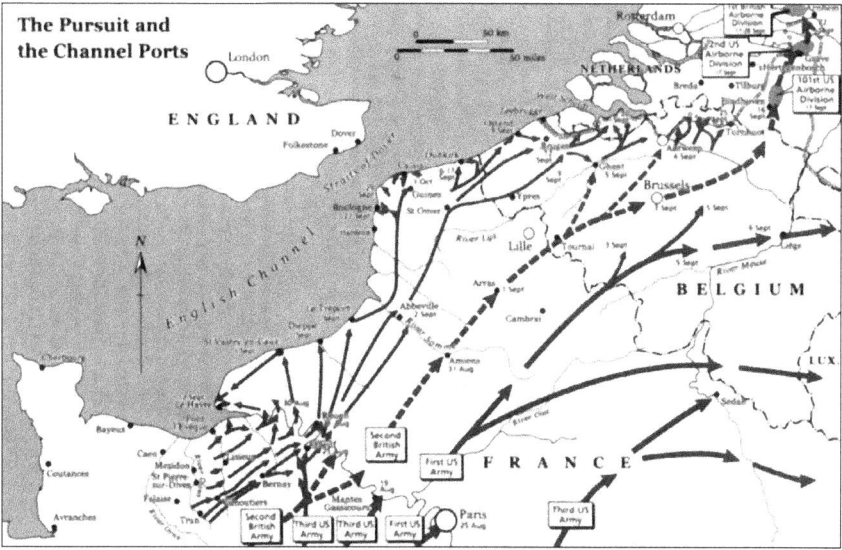

'The Pursuit and Channel Ports.' By the end of 1944, the British Liberation Army had reached the frontiers of the Third Reich.

Architects of victory in Operation Eclipse: Field Marshal Bernard Montgomery with Major General Adair (Guards Armoured) left, Lt General Horrocks (30 Corps Commander) right and Major General 'Pip' Roberts (11th Black Bull Armoured) extreme right. (*Imperial War Museum*)

Foreword

This book is the sequel to *The Rhine Endeavour*, which documented the events of September 1944, when the British Liberation Army liberated 20 million of Adolf Hitler's victims – French, Belgian and Dutch. Poised on the borders of the Third Reich, within five months a further 20 million civilians of Holland, Norway, and Denmark, and a multitude of Allied prisoners of war and Displaced Persons (DP), were rescued in Operation Eclipse.

The British Liberation Army brought fearful war, peace, justice and retribution. This is the story of Monty's young soldiers – Patrick Delaforce was one of them – who helped end the ignoble Third Reich.

The author (on right), troop leader with 13 RHA in Kiel, May 1945, who fought in the five major battles in Operation Eclipse.

The British Liberation Army

This is the story of Operation Eclipse – the grand design – the plan for the British Liberation Army (BLA), as part of the Allied Command (Americans, Canadians, French and Poles), to invade and conquer the Third Reich in 1945.

By the end of 1944 much had been achieved. Operation Overlord, the invasion into German-defended Normandy, after eleven weeks of attritional fighting during which battlefield casualties averaged 7,000 a day, had been an outstanding success. Hitler had sent his best panzer troops, SS divisions and Hitler Jugend in and they were outnumbered, outgunned, smashed by the Allied air forces, pulverised by the massed artillery and not even the Normandy *bocage* defensive terrain could save them.

The Great Swan followed. General 'Blood and Guts' Patton raced through Brittany and headed for Paris. Very politely, General de Gaulle's French troops were allowed in first. In Operation Anvil dozens more American divisions swept up the Rhône valley to strengthen Eisenhower's huge Allied army.

Next came the British Great Swan, when General Montgomery's four armoured divisions, 7th Desert Rats, 11th Black Bull, the Guards and the Poles, swept north for 250 miles through western and northern France into the Low Countries.

September 1944 was the 'zenith' month for the BLA. Le Havre was captured in a classic all-services battle in Operation Astonia. The Desert Rats and the Poles liberated Ghent, Lille and all the northern French towns; famous names of the Great War all fell. The Guards Armoured thrust into Brussels for the untold, unimaginable joy of liberation. The 11th Armoured Division, in which the author served, with a lot of help

Field Marshal Bernard Montgomery led the British Liberation Army to victory in Operation Eclipse. (*Imperial War Museum*)

from the Belgian Liberation Army, captured the greatest prize of all – Antwerp city and its huge, great dockland.

It was too good to be true!

Euphoria reigned. All the field marshals and generals joined in a litany of great expectation: 'War will be over by Christmas,' 'On to Berlin.' The liberated citizens of France, Belgium and some in Holland said of their German conquerors, 'Hitler kaput.'

In the autumn of 1944 Alan Moorehead, the intrepid Australian war journalist, wrote:

> It was a most dangerous period of delay. Every hour, every day, the German morale was hardening. As the broken remnants of the Fifteenth and Seventh Armies struggled back to the Reich they were regrouped into new formations [kampfgruppen]. Anything and everything served at this desperate moment. Submarine crews were put in the line as infantry; the German water-police were mobilised; there was a brigade of deaf men who presumably received their orders in deaf and dumb language. There was a whole division of men who suffered from stomach ailments and had to be served with special bread.

Throughout the Reich every officer and man on leave was summoned back to his post. Extraordinary prisoners were collected – near-sighted clerks who had left their city offices three weeks before; men with half-healed wounds, even cripples and children of fifteen or sixteen [about to join the dreaded Hitler Jugend]. It was a makeshift hotch-potch army, an emergency army put in simply to hold the gap, simply to fight for time while the German generals reorganised on a sounder basis. Little by little a crust was formed along the valley of the Rhine from the Swiss border to the Zuyder Zee.

On 6 December 1944 Prime Minister Churchill wrote to President Roosevelt:

> The time has come for me to place before you the serious and disappointing war situation which faces us at the close of the year ... The fact remains that we have definitely failed to achieve the strategic object which we gave our armies five weeks ago. We have not reached the Rhine in the Northern part and most important sector of the front. We shall have to continue the great battle for many weeks before we can hope to reach the Rhine and establish our bridgeheads. After that again, we have to advance through Germany.

By the beginning of December 1944, the official statistics for the rosters of the German Feldheer (armed forces strength) showed over 10 million soldiers able and probably willing to fight and die for their Führer. This included the Luftwaffe, Navy, coastguards, police, the new Volks-grenadier divisions, the Volksturm (Home Guard), convalescent units, sweepings from the German prisons, multitudes of 'pressed' men or 'volunteers' from the east (Ost-truppen), Poles, Czechs, Rumanians, Russians and the rather dangerous SS 'foreign' turncoats from the occupied territories. They all joined in – France, Belgium, Holland, Italy, and the Vikings from Scandinavia – providing young, dedicated men who, as renegades, knew what their future would be if captured. Probably there would have been a British SS unit or two if Operation Sealion had been successful.

The comparative failure of Operation Market Garden in September had been caused by many factors. Undoubtedly the main cause was the astonishing skill and resolution of the German field commanders who marshalled aggressive defences to thwart Field Marshal (from 1 September) Bernard Montgomery's huge army thrusting into the centre of Holland.

Hitler's generals – Model (left) confers with Student, Bittrich and Harmel.

Liddell Hart, the distinguished military expert, interviewed many of Hitler's commanders immediately after the end of the war and wrote: 'The German generals of this war were the best-finished product of their profession – anywhere. They could have been better if their outlook had been wider and their understanding deeper. But if they had become philosophers they would have ceased to be soldiers.' Any student of history who reads Field Marshal Michael Carver's comments in his memoirs on the many British generals he served with or under will agree. The German generals were far better than the vast majority of the Allied generals. Sad, but true!

Although some of Hitler's *festung* fortress commanders were mediocre, most of them defended their ports with great tenacity. Indeed, Dunkirk was so well protected that it was deemed pointless to grind the defenders down and they eventually surrendered at the end of the war – totally undefeated! Old von Rundstedt, certainly Field Marshal Model ('Hitler's Fireman'), Bittrich and Student, who commanded the Parachute Army, were probably more experienced and capable than most of the Allied generals. General von Zangen, for example, created a German 'Dunkirk'

Monty and his generals: Dempsey (on his left), Crerar (on his right), Thomas (on his far right) and 'Pip' Roberts (behind his right shoulder). Behind Monty are Scottish generals wearing Tam o'Shanters.

by skilfully extracting the bulk of the 15th Army from the Pas de Calais north and across the river Scheldt, at night, not only into safety but to add to the German defences in Holland. He wrote:

> Therefore, I order commanders, as well as the National Socialist indoctrination officers, to instruct the troops in the clearest and most factual manner on the following points: Next to Hamburg, Antwerp is the largest port in Europe …
>
> After overrunning the Scheldt fortifications, the English would finally be in a position to land great masses of material in a large and completely protected harbour. With this material they might deliver a death blow to the north German plateau and to Berlin before the onset of winter.
>
> In order to pretend that the battle of Antwerp is unimportant and to shake the morale of its defenders, enemy propaganda maintains that the Anglo-American forces already possess sufficient ports which are intact, with the result that they are not at all dependent on Antwerp. That is a known lie. In particular, the port of Ostend, of which the enemy radio speaks frequently, is completely destroyed. Current delays in the enemy's conduct of the war are attributable in great measure to the fact that he still must bring all his supplies through the improvised facilities of Cherbourg. As a result, he has even had to lay a temporary oil pipe-line from these to the interior of France …

In his last speech, Churchill said again, 'before the storms of winter we must bring in the harvest in Europe'. The enemy knows that he must assault the European fortress as speedily as possible before its inner lines of resistance are built up and occupied by new divisions. For this reason he needs the Antwerp harbor. And for this reason, we must hold the Scheldt fortifications to the end. The German people is watching us. In this hour, the fortifications along the Scheldt occupy a role which is decisive for the future of our people. Each additional day that you deny the port of Antwerp to the enemy and to the resources that he has at his disposal, will be vital.

(Signed) VON ZANGEN
General of the Infantry

Hitler, Göring and Göbbels had between them conjured up a huge army despite the staggering losses on the Russian front, in North Africa, and in Normandy, which still numbered over ten million men. Göring's Parachute Army from Luftwaffe sources was young, fanatical and seemed happy to fight and die for the Third Reich. It was deployed immediately opposite the BLA, and in the late autumn of 1944 often won the day. The SS troops were the *crème de la crème* of Hitler's forces, but unfortunately they left a long trail of atrocities behind them. The author saw one at Breendonk fortress south of Antwerp, where the Belgian SS and German Gestapo had tortured, maimed and executed anyone who might be part of the Resistance and, of course, the Jewish population.

Other disquieting factors were that the German armies had recovered from Normandy and were fighting in hastily assembled kampfgruppen in the polders and Peel country of Holland with great vigour and audacity. By contrast the BLA, with limited sources of manpower due to its immense losses of mainly 'Poor Bloody Infantry' (PBI) in Normandy, had had to disband over twenty formations, including the magnificent 50th Tyne-Tees Northumbrian Division. As the Luftwaffe was in almost total decline, the need for the anti-aircraft regiments in every one of Montgomery's divisions was now deemed not to be essential. Many of the anti-tank artillery regiments in every division were less vital, as Tigers, Panthers, Leopards and the host of Teutonic panzers had suffered immense losses. And the survivors were fighting desperately on the Eastern Front to keep Stalin's hordes at bay, so tens of thousands of young men in khaki swapped jobs for a much more dangerous one and joined 3rd British, 15th Scottish, 43rd Wessex, 49th Polar Bears,

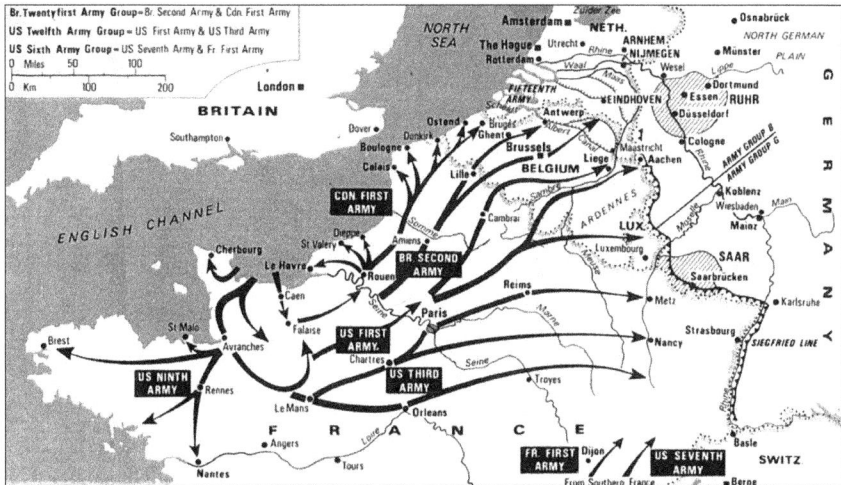

Battlegrounds of France, Belgium, Holland, into Germany.

51st Highland and 53rd Welsh divisions who were desperate for reinforcements.

The balance of power between the Allies had been swinging inevitably away from Prime Minister Churchill and the British contribution to winning the war. True, the Royal Navy and the RAF were performing miracles practically every single day, but on land it was a different matter.

On the Eastern Front 410 Russian and German divisions were battering each other with staggering losses on both sides. But on the Western Front 'only' 120 divisions were locked in combat. At the end of 1944, Hitler had 133 divisions at war in Russia, 76 in north-west Europe, 24 in Italy, 10 in Yugoslavia and 17 lurking in the wings in Norway and Denmark.

General Eisenhower then had a total of 79 US and French divisions under his command (49 infantry, 26 armoured and 4 airborne). Field Marshal Montgomery had 12 British, 3–5 Canadian, some who had arrived from Italy, and one Polish division.

At their few 'Great Power' meetings, Churchill's influence was waning, President Roosevelt was ailing and was soon to die, and Josef Stalin made promises without intending to honour them. His agenda was quite straightforward: bloody revenge on Hitler's wicked Operation Barbarossa and the maximum of communist influence and power in the

Monty (second from right) and his ADCs in front of three war caravans.

Baltic States and Eastern Germany, including the capture of Berlin and poor Poland, for whom many years back – or so it seemed at the time – Britain and France went to war.

Josef Stalin stated cynically, but accurately: 'England provided the time, America provided the money and Russia provided the blood.' With a huge population of 190 million, against the 80 million of the Third Reich, it was clear that Stalin could accept huge losses of men and territory until, with much help from 'General Winter', the tide turned. The failure to take Moscow and Stalingrad and the battles of Kiev and Kursk would soon prove decisive. In the Second World War, Russian military losses were about 11 million, and Nazi losses 3.8 million. Even worse, about 23–26 million Soviet citizens died, one way or another, during the war. Millions more endured wounds, hardship, loss of homes and land, and captivity in Germany.

By contrast the USA service deaths in all theatres of war numbered 405,399 and those of the British numbered about 190,000.

There were still terrible tribulations lying ahead of the BLA: the thunderbolt in the Ardennes, and operations Blackcock, Veritable, Blockbuster, Anger, Varsity and Plunder. There were painful landmarks

16

21 ARMY GROUP

PERSONAL CHRISTMAS MESSAGE
FROM THE C-IN-C

(To be read out to all troops)

1. The forces of the British Empire in western Europe spend Christmas 1944 in the field. But what a change has come over the scene since last Christmas.

The supreme Battle of Normandy carried with it the liberation of France and Belgium.

Last Christmas we were in England, expectant and full of hope; this Christmas we are fighting in Germany.

The conquest of Germany remains.

2. It would have needed a brave man to say on D day, 6 June, that in three months we would be in Brussels and Antwerp: having liberated nearly the whole of France and Belgium; and in six months we would be fighting in Germany: having driven the enemy back across his own frontiers.

But this is what has happened.

And we must not fail to give the praise and honour where it is due:

"*This was the Lord's doing, and it is marvellous in our eyes.*"

3. At Christmas time, whether in our homes or fighting in the field, we like to sing the carols we learnt as children; and in truth, this is indeed a link between us and our families and friends in the home country: since they are singing the same verses. The old words express exactly what we all feel today:

"*Glory to God in the highest, and on earth peace, good will toward men.*"

That is what we are fighting for, that is what we desire: on earth peace, good will toward men.

4. And so today we sing the Christmas hymns, full of hope, and steadfast in our belief that soon we shall achieve our hearts' desire.

Therefore, with faith in God, and with enthusiasm for our cause and for the day of battle, let us continue the contest with stout hearts and with determination to conquer.

5. And at this time I send to each one of you my best wishes and my Christmas greetings.

Wherever you may be, fighting in the front line, or working on the lines of communication or in the ports, I wish all of you good luck and a happy 1945. We are all one great team; together, you and I, we have achieved much; and together, we will see the thing through to the end.

6. Good luck to you all.

B. L., Montgomery
Field . Marshal
C-in-C 21 Army Group.

Belgium.
Xmas, 1944

Monty's Christmas message.

on the way. The appalling concentration camp of Bergen Belsen, the difficult battle to capture Bremen, and the seizure of ruined Hamburg, but among all the trauma of daily armed confrontation was the fact that the BLA succeeded in bringing peace and liberation to nearly 20 million of Hitler's victims: most of Holland, aided by the Canadians and Poles; the sovereign countries of Denmark and Norway; the liberation of vast numbers of prisoners of war; and several million (out of six million) nomadic displaced persons. These slaves of the Hitler regime worked in the factories, on the farms, and in the defences of Germany. Poor devils – many of them knew not how or where to go. The many Russian 'serfs' faced a dangerous future if they returned to their native country. But they were, or would be by May 1945, freed from Nazi captivity.

Unbeknown to the Allied leaders on New Year's Day 1945, there were three great moral hazards ahead of them. Chapter 25 chronicles the long, unhappy final year, when most of the Dutch citizens were starving. Seyss-Inquhart, the German Reichskommissar for the Netherlands, deliberately created the *Hongerwinter* of 1944–45. The Dutch royal family and government in exile pleaded for the Allies, who in Market Garden had fought and controlled a sizeable part of the country, to finish the job. But Seyss-Inquhart was already destroying polder-dykes and flooding prosperous sectors and was threatening even more. And Field Marshal Montgomery and Eisenhower too, decided that the whole purpose of Operation Eclipse – the destruction of the Third Reich – was paramount. So Holland starved! Operation Manna, a humanitarian food relief programme, was finally agreed in late April 1945.

Among the millions of displaced persons roaming Germany during 1945 were over two million captured Russian prisoners of war. On release many of them were frightened of being repatriated; to be a prisoner of war in the peculiar Russian mindset was considered to be treasonable.

The Allied commanders in the West were confronted with this terrible dilemma. There was no reason to believe that, on their return, the vast majority would be sent arbitrarily to the hated *gulags*. The Geneva Convention rules were that third party prisoners of war were to be returned to their homeland.

When finally the Allied Commission had to organise some stable law and order in the defeated Third Reich, with every major city and town destroyed, the best candidates able to obtain order out of chaos were all loyal, dedicated Nazis, although many of course denied that. The millions of German prisoners of war were 'screened'. The vast majority were released in a short time, a small minority were charged with war crimes and tried, and another small minority, including a future president, were allowed to continue supervising their fellow countrymen, albeit peacefully. It was difficult to imagine the same thing happening if Operation Sealion had succeeded. Britain would have been under the jackboot and all their leaders permanently in prison camps or dead and buried!

Retribution was also a moral issue. Proven war criminals should face justice in appropriate tribunals under international law, which should, above all, be seen to be eminently fair. Nuremberg was unique in its scale and was observed by the press of the world. Justice was seen to be done. There were also minor war crimes tribunals in Hamburg and Luneberg, which are covered in this book. The author was a junior officer in 3rd RHA, Desert Rats, at the time and was one of three officers who combined jury and judicial duties, advised by an eminent Judge Advocate General. In Hamburg, several dozen 'minor' war criminals, mostly prison guards, were tried and sentenced. Later, just before his twenty-second birthday, he witnessed, as one of fifty official British Army of the Rhine (BAOR) officers, Mr Albert Pierrepoint, the official British hangman, in action in Hameln.

This book is about the way in which the citizen-army of Great Britain, the BLA, emerged from the slough of despondency in Holland, smashed, slowly and very painfully, through the Siegfried Line, defeated Hitler's armies west of the Rhine, and fought five river crossing battles to reach the Baltic just before Stalin's Cossack army marauders. On the way they stormed the vital port of Bremen and captured Hamburg.

They brought peace and liberation to most of Holland, to Denmark, to Norway, to tens of thousands of Allied prisoners of war, and succour for some of the millions of nomadic homeless and displaced persons, effectively slaves of the Nazi regime. The BLA forced a corridor about 100 miles wide and thrust ahead some 250 miles. The German army claimed in 1918/19 that they had not been defeated, as at the Armistice they withdrew to the Fatherland. The author's father fought them in Flanders fields and then became a member of the BAOR. His son did exactly the same thing some twenty-five years later. The main difference was that for the first time ever the German country was invaded and conquered, 'Alles kaput.'

Above: Liberation.

Left: Freedom.

Below: Retribution.

CHAPTER ONE

Hitler's Crumbling Third Reich

One of the greatest mysteries of the twentieth century was how a motley collection of not very intelligent German street brawlers, led and totally dominated by one man, could take most of Europe and threaten the rest of the world with brutal war. Adolf Hitler (1889–1945), an Austrian misfit and loner, with a slight education, no family genes of note and no political background, created the infamous Third Reich by sheer determination, confidence, willpower, deviousness and luck. It was incredible that a brave, undistinguished Great War corporal, who never fired a gun in anger, and was distinctly not officer material, should encourage, organise and control the world's most brilliant fighting force, and control them from one of his twenty headquarters until the sordid end.

William Shirer was a determined American CBS journalist resident in Berlin in the late 1930s and into the start of the Second World War who witnessed the Führer's activities for several years. He attended a colossal Nuremberg rally for the Nazi faithful. There were 30,000 people in the great Luitpold Hall, tens of thousands packing the narrow streets to get a glimpse of Hitler and watch the goose-stepping military might, and half a million at the Zeppelin Meadow. Shirer wrote:

> You have to go through one of these to understand Hitler's hold on the people, to feel the dynamism in the movement he's unleashed and the sheer, disciplined strength the Germans possess. And now – as Hitler told the correspondents yesterday – the half-million men and women who've been here during the week will go back to their towns and villages and preach the new faith with new fanaticism.

Hitler rewarding his youth troops. Hitler 'Jugend' in the making!

Back in Berlin I tried to sort over my first impressions of Adolf Hitler. In looks and appearance he was rather different from what I had expected. In Nuremberg, except during the traditional hysterical outbreak against the Bolsheviks and the Jews, he did not look or behave like the mad, raging, brutish dictator I had been reading about and which he had proved himself to be by his merciless massacre of the SA leaders and political opponents less than three months before.

His face was rather common. It was coarse. It was not particularly strong. Sometimes, when he obviously was fatigued from the long speeches, the hours spent in reviewing his troops, it appeared flabby.

All through the week in Nuremberg, where I often sat or stood but a few feet from him in the stands, I tried to size up the features of the man. He was about five feet, nine inches tall and weighed around 150 pounds. His legs were short and his knees turned in slightly so that he seemed to be a bit knock-kneed. He had well-formed hands, with long, graceful fingers that reminded one of those of a concert pianist, and he used them effectively, I thought, in his gestures during a speech or when talking informally with a small group.

His nose betrayed the brutal side of him. It was straight but rather large and broadened at the base, where thick nostrils widened it. It was the coarsest feature of the face. Perhaps it was to soften it that he grew his famous Charlie

Chaplin mustache. His mouth was quite expressive, and it could reflect a variety of moods.

It was the eyes that dominated the otherwise common face. They were hypnotic. Piercing. Penetrating. As far as I could tell, they were light blue, but the color was not the thing you noticed. What hit you at once was their power. They stared at you. They stared through you. They seemed to immobilize the person on whom they were directed, frightening some and fascinating others, especially women, but dominating them in any case. They reminded me of paintings I had seen of the Medusa, whose stare was said to turn men into stone or reduce them to impotence. All through the days at Nuremberg I would observe hardened old party leaders, who had spent years in the company of Hitler, freeze as he paused to talk to one or the other of them, hypnotized by his penetrating glare. I thought at first that only Germans reacted in this manner. But one day at a reception for foreign diplomats I noticed one envoy after another apparently succumbing to the famous eyes. Martha Dodd, the vivacious young daughter of the American ambassador, had told me a day or two before I left for Nuremberg, to watch out for Hitler's eyes. 'They are unforgettable,' she said. 'They overwhelm you.'

His oratory also was overwhelming. at least to Germans. It held them spellbound. At Nuremberg I grasped for the first time that it was Hitler's eloquence, his astonishing ability to move a German audience by speech, that more than anything else had swept him from oblivion to power as dictator and seemed likely to keep him there.

The words he uttered, the thoughts he expressed, often seemed to me ridiculous, but that week in Nuremberg I began to comprehend that it did not matter so much what he said but how he said it. Hitler's communication with his audiences was uncanny.

Hitler was a great gambler, not for money but for the highest stakes of war. In 1940 and 1941, every gamble came off: brave Poland was over-run in weeks; at Munich Hitler had walked all over Neville Chamberlain; 'Weser Exercise' swallowed up Denmark and Norway; 'Fall Gelb' swamped Holland, Belgium and France when Europe's largest army simply imploded and the famous Maginot Line was bypassed at both ends.

In December 1940 William Shirer in Berlin wrote: 'When I left, German troops, after their quick and easy conquests of Poland, Denmark, Norway, Holland, Belgium and France, stood watch from the North Cape to the Pyrenees, from the Atlantic to beyond the Vistula. Britain stood alone. Few doubted that Hitler would emerge from the conflict as the greatest conqueror since Napoleon. Not many believed that Britain would survive.' Soon Yugoslavia and Greece followed and were over-run.

The great gambler had called 'heads' every single time and had guessed right. But his luck started to run out. 'Tails' it was as Rommel's German and Italian armies surrendered in North Africa. The aerial battle over Britain had not been won.

Soon hubris set in. The world lay at the Führer's feet. His well-trained Wehrmacht and Luftwaffe were all-conquering. In his book *Mein Kampf* he had spelt out his objectives for all the democracies to read. Of course, the Führer, the architect of the Third Reich, which would last for a thousand years, was locked into his own private cuckoo-land of make-believe unblemished by reality. On every front – Italy, Russia and north-west Europe – his ten million strong warriors were in disarray.

If he had not committed two acts of lunacy, perhaps three, the world, and certainly Europe, would have had an unprecedented era of despair, violence and little hope. Operation Barbarossa was utter madness. The evil Joseph Stalin had absolutely no intention of invading Germany – for all sorts of excellent reasons. The second act – vainglorious and stupid in spades – was the declaration of war against the United States to show some unnecessary solidarity with his far-flung allies in Japan. President Roosevelt had spent several years being converted, persuaded and cajoled by Winston Churchill into stepping up 'commercial' help (such as fifty decayed, obsolete destroyers), to help and incidentally, bankrupt his distant ally. A formal declaration of war was what Churchill desperately needed, but Roosevelt's voting public would never have allowed it.

These were the two obvious mistakes. The third and perhaps just as deadly is shown in Appendix B: 'Hitler's Nuclear Option.' If Hitler had not been told that all his very, very eminent nuclear physicists were Jewish, and therefore 'sub rosa' unreliable and untrustworthy, his race for the ultimate vengeance weapon could, maybe would, have been achieved.

At the Führer's naval conference on 8 July 1943, Grand Admiral Dönitz advised Hitler that two new types of submarine had been designed, which would revolutionise the war at sea. These new *Die Boot* could travel as fast *under* water as on the surface. They could operate submerged for prolonged periods without having to come up to charge their batteries. Dönitz said: 'Entirely new possibilities are introduced by permitting U-boats to approach a convoy quickly. They could also take swift evasive action at 15 knots underwater instead of being obliged to surface. This will make the enemy's present anti-submarine defence entirely ineffectual.' Dönitz was delighted. Hitler was delighted and Albert Speer was told to grant top priority to the construction of these new U-boats. Dönitz's team reckoned a minimum of 17 months was needed. Speer was confident that by not building prototypes and testing them, he could reduce the time required by half. By December 1943 a full-scale mock-up of each model had been built and the first electro U-boats would be ready by the spring of 1944.

Despite providing his existing fleet with better defensive radar, heavier anti-aircraft armament and a new acoustically guided torpedo, Dönitz knew that for every ship his U-boats sank the Americans were building four new Liberty vessels. The Allied long-range air patrols were becoming more effective and the British Royal Navy convoy systems were better protected.

Allied bombing caused delay to Speer's production. A Type XXI prototype was destroyed in Kiel harbour and many of the subcontractors lacked the necessary skill and precision required. Even the new Schnorkel, which allow batteries to be recharged without surfacing, had many technical problems. Nevertheless, in mid-1944 Dönitz still had well over 400 serviceable U-boats.

By early 1945, 128 new electro U-boats, not the 245 planned, were ready for sea trials. Bletchley Park's Ultra machine had detected clues about Dönitz's new dangerous 'toys', so from the end of 1943 the construction yards had been bombed and the shallow waters of the Baltic had been mined.

The famous war correspondent, Chester Wilmot, pointed out in his majestic *Struggle for Europe* that because Grand Admiral Dönitz was pushing the claims of his new electro U-boats, which required the deeper waters of the Gulf of Danzig, Memel, East Prussia and the Courland Pocket in western Latvia for their trials, Hitler felt compelled to leave forty *vital* Wehrmacht divisions to defend the eastern Baltic shores from Konigsberg northwards from the Russian Red Army. Even though in February 1945 Dönitz had 450 U-boats – 'the largest number that Germany had ever possessed', he told the Führer – the formidable thrust by the BLA reached Kiel harbour in time to ensure that the electro U-boats and two-men midget submarines were captured.

So Hitler's dreams of lethal new victory weapons, V-1s, V-2s, the ME jet fighters, the new U-boats, and his half-hearted efforts with an atomic

Battered ruins of Cleves.

bomb, had all failed. They would have been nightmares – fortunately, they were mainly thwarted by the interdiction by the Allied air forces.

The Führer must have realised by February that his vaunted Vengeance weapon [*vergeltungswaffe*], conducted by Flak Regiment 155, had not brought Britain to its knees. 15,000 V-1s, the noisy, pilotless and grossly inaccurate monoplane, which was targeted on London and Antwerp, had caused terror on a limited scale. About 6,000 of them had not reached their targets. Next came the V-2, a liquid-fuelled rocket travelling at supersonic speed, which was impossible to intercept, unlike the V-1 which could be shot down or even gently nudged. The 5,000 V-2s with the same one-ton high explosive warhead failed to break the British spirit. Indeed, there was evidence that Bomber Harris's obliteration of every major town in Germany had failed to break the German collective will to force a regime change.

Albert Speer had worked production miracles in the battered Third Reich. Hitler made his architect friend Armaments Minister in 1942, and he became second only to Hitler himself as a power on the home front. Skoda factories at Pilsen produced panzer tanks. Two million French POWs worked in the mines and on the farms to release German manpower for the armed services. A further 7.8 million slave labourers were made to work in the Ruhr. Factories went underground to escape the bombing or were rebuilt outside the towns. Oil came into the Reich from Romania. Synthetic oil production, despite Bomber Harris' 1,000 bomber raids on the target, remained steady at between 4 and 5 million tons each year, peaking at 5.7 million in 1943. Tank production rose from 6,000 in 1942, to 12,000 in 1943. Averaging over 30 million tons in 1941, coal production averaged over 250 million tons in 1944. It was incredible!

Hitler ruled his generals and *festung* fortress commanders with an iron fist. Even portly Field Marshal Hermann Göring, the Luftwaffe boss and nominally No. 2 in the Nazi hierarchy, was petrified of Hitler. Göring's promises to destroy the British RAF in Operation Sealion, which had ended in dismal failure, meant that he was never fully trusted again. Several field marshals committed suicide. Operation Valkyrie, the assassination plot in July 1944, had made Hitler even more paranoid. His remaining top generals, Model, von Rundstedt, Kesselring, Guderian and Kettel stayed faithful even though they knew that many of the orders they received were impossible to carry out. They tried their best.

First 'Hitler's Fireman', Field Marshal Model:

14 October 1944
ORDER OF THE DAY
Soldiers of the Army Group!
The battle in the West has reached its peak. On widely separated fronts we must defend the soil of our German homeland. Now we must shield the sacred soil of the Fatherland with tenacity and doggedness … *The Commandment of the hour is: None of us gives up a square foot of German soil while still alive.*

Every bunker, every block of houses in a German town, every German village must become a fortress which shatters the enemy. That's what the Führer, the people and our dead comrades expect from us. *The enemy shall know that there is no road into the heart of the Reich except over our dead bodies …*

Egotism, neglect of duty, defeatism and especially cowardice must not be allowed room in our hearts. *Whoever retreats without giving battle is a traitor to his people …*

Soldiers! Our homeland, the lives of our wives and children are at stake!

Our Führer and our loved ones have confidence in their soldiers! We will show ourselves worthy of their confidence.

Long live our Germany and our beloved Führer!
MODEL
Field Marshal

Then there followed another message to the troops:

I expect you to defend Germany's sacred soil with all your strength and to the very last. The homeland will thank you through untiring efforts and will be proud of you.

New soldiers will arrive at the Western front. Instil into them your will to victory and your battle experience. All officers and NCOs are responsible for all troops being at all times conscious of their great responsibility as defenders of the Western approaches. Soldiers of the Western Front!

Every attempt of the enemy to break into our Fatherland will fail because of your unshakeable bearing:
Heil the Führer!
VON RUNDSTEDT
Field Marshal

So through the combined efforts, threats, and bullying by Hitler's formidable coterie, Himmler, Göbbels and Göring, the strength towards the end of 1944 was Wehrmacht and Waffen SS 7,536,946; Luftwaffe 1,925,29; and Kriegsmarines 703,066 – a total of 10,165,303.

Heinrich Himmler, Hitler's terrible henchman, who organised the 'Final Solution', the supervision of the destruction of Jewry, not only in the Reich but also in every occupied country, had taken over command of the 'Home Army' in Germany after the failed assassination attempt on 20 July.

In the autumn of 1944, a document called 'Enemy War Aims' was officially propagated across the Third Reich by Himmler.

In May 1944 the Soviet Union offered a plan to her Allies whereby all members of the German forces would be made prisoners-of-war and organised into labour battalions for forced labour in the Soviet Union. The USA as well as England agreed to the proposal without restrictions.

On top of that, the USA demanded that the geographical unity of Germany be dissolved by running corridors through it, which would be inhabited by non-Germans. They intend to distribute these sectors to 'a population with more peaceful tendencies' while making Germany defenceless and enslaved.

To destroy German national pride, a complete understanding exists between Moscow, London and Washington.

(a) They will not make peace with Germany, but will continuously occupy the country.

(b) They intend to intermarry Germany women with members of the occupying forces.

(c) German children will be educated after separation from their parents ...

The will to destroy is not directed against National Socialism but, without any doubt, against the entire German nation.

Whoever does not fulfil his duty to the utmost during the coming weeks and months is helping to make these war aims of the enemy a terrible reality.

And to his SS troops Himmler sent this brutal message:

Reichsführer SS 10 September 1944

Certain unreliable elements seem to believe that the war will be over for them as soon as they surrender to the enemy.

Against this belief it must be pointed out that every deserter will be prosecuted and will find his just punishment. Furthermore his ignominious behaviour will entail the most severe consequences for his family. Upon examination of the circumstances they will be summarily shot.

(Signed) HIMMLER

Many thousands of the Reich SS had been forced to surrender in Normandy. The survivors were now warned.

The arch-propagandist, Dr Joseph Göbbels, had a very difficult task – on every front in late 1944 the great armies of 1940, 1941, 1942 and 1943 were in full retreat. The homeland had been bombed to destruction, the enemy were nearing the borders. Milton Shulman, who interviewed most of the key German commanders immediately after the end of the war, wrote:

And when the commanders had paused for breath, Göbbels and his propaganda boys stepped in to continue pumping into the dazed and weary body of the German soldier the synthetic stimulants of fear and faith which alone could keep him resisting. But the propaganda line had undergone a change since the early days of the invasion. Then the emphasis had been on hope – the promise of secret weapons and secret armies which would suddenly appear to crush the Allies in one violent cataclysm. The secret weapons had come and gone and yet the shadow of defeat was closer and darker than it had ever been before. The promise of secret weapons now had to be soft-pedalled. The emphasis was now on fear – fear of the destruction of Germany, of the rape of its womanhood, of the vengeance of the Russian Bolsheviks. Hammering on these themes like some mad musician at an organ, Göbbels pulled out all the stops, reaching a Wagnerian crescendo of frenzied hysteria. One typical example of the propaganda efforts used to frighten or shame the German soldier into staying in the line was a leaflet entitled 'The Watch on the Rhine'. Over the picture of a medieval German warrior, complete with flowing cloak, chain-armour and enormous sword, standing guard on the Rhine, the following words, amongst others, were written:

Comrade, the enemy means to outflank the West Wall at the very point where we are and to cross the Rhine into Germany!

Shall our people, shall our families, have suffered five years in vain?

Shall they suffer misery and starvation amid the ruins of our cities in a conquered Germany?

Do you wish to go to Siberia to work as a slave?

What do you say about it?

Never shall this happen.

Never shall the heroic sacrifices of our people prove in vain!

Therefore, everything depends now on your courage! The struggle against an enemy who at the moment is still superior is tremendously hard. But for all of us there is no other way out than *to fight on with knives if need be.*

It is better to die than to accept dishonour and slavery!

It is better to be dead than a slave!

Therefore – keep the watch on the Rhine steadfastly and loyally.

Nevertheless, Hitler's Third Reich was crumbling!

The vast majority of the BLA poised on the frontier of the Third Reich – poised to break and enter a strange, hostile land – had never visited Germany before, and almost certainly had never met any Germans socially. The front line soldiers had encountered the SS and the Hitler Jugend and did not like what they saw. They and their families had suffered almost five years of total war, perhaps in the many towns blitzed by the Luftwaffe. They thought they knew quite a lot about Hitler, his revolting gang and 'Lord Haw-Haw', the renegade British radio 'comic'. Almost certainly they knew little about the appalling butchery of the Jews (and many others) in the extermination and concentration camps.

Alan Moorehead wrote in *Eclipse*:

> For most of us it was our first experience of Germany itself, the first time we had come into contact with large numbers of German civilians and seen the way they lived and dressed and carried themselves. It was the beginning of an immensely complicated relationship between ourselves and the defeated, a story that kept changing its plot ... As soon as you discovered evil and malice in one place you were immediately confronted with kindness and genuine innocence in another ... No matter where you went, or what you did, was placed against the unending tragedy and physical ruin of the country.

The black, brutal, sinister side of the Third Reich was hidden from the vast majority of the BLA and from most of the journalists. The appalling atrocities enacted at Auschwitz, Buchenwald, Belsen and Dachau were only discovered in the last few dying weeks of the war. In the excellent book *War Report, Despatches: D-Day to VE Day*, BBC war correspondents – Robert Reid on Buchenwald, Richard Dimbleby on Belsen, and Ian Wilson on Dachau – write about their firsthand on-the-spot visits to those hells on earth. The appendix on the camps at the end of this book is not for the faint-hearted. It is the testimony to the lasting shame of Hitler's Third Reich.

Total war – destruction of the Third Reich.

CHAPTER 2

Making War

Ahead for the BLA lay several set-piece 'Monty' attacks planned down to the last detail. Overwhelming close air support (CAS) by Typhoon cab ranks and colossal barrages, and under that shelter – a hideous, noisy umbrella – infantry supported by tanks, flails, Crocodiles and petard-throwing AVREs would make their slow, very careful advance.

However, it often failed to work out quite as planned!

The War Office's 'Current reports from Overseas' secured reports made at the time by eminent German military commanders. One from the Italian front noted:

> The conduct of the battle by the Americans and English was, taken all round, once again very methodical. Local successes were seldom exploited ... British attacking formations were split up into large numbers of assault squads commanded by officers. NCOs were rarely in the 'big picture', so that if the officer became a casualty, they were unable to act in accordance with the main plan. The result was that in a quickly changing situation *the junior commanders showed insufficient flexibility*. For instance, when an objective was reached, the enemy would neglect to exploit and dig in for defence. The conclusion is: as far as possible *go for the enemy officers*. Then seize the initiative yourself.

The Panzer Lehr, the finest German armoured division in Normandy, under General Fritz Bayerlein, in an intelligence report noted that:

> ... a successful break-in by the enemy was seldom exploited in pursuit. If our own troops were ready near the front for a local counter-attack, the ground was immediately regained. Enemy infantry offensive action by night is limited

33

South Lancashires advance on Venray during 'Operation Aintree', November 1944. *(Imperial War Museum)*

Infantry of 3rd British Division move up. Mine sweeping on the right, tank support on left. (*Imperial War Museum*)

to small reconnaissance patrols ... It is better to attack the English who are very sensitive to close combat and flank attack at their weakest moment – that is, when they have to fight without their artillery.

A British corps commander in the Italian campaign wrote:

> The destruction of the enemy was most easily achieved when we managed to keep him tired and in a state of disorganisation, which resulted in unco-ordinated defence and lack of food, petrol and ammunition. We were undoubtedly too inelastic in our methods when faced with changing conditions. After six weeks of mobile fighting, during which the enemy never launched anything bigger than weak company counter-attacks, we still talked too much about 'firm bases' and 'exposed flanks'.

Time and time again the German defenders held their front line with a minimum of strength, but with many observation posts to bring down heavy defensive nebelwerfer 'stonks'. The thin defensive line protected by substantial minefields absorbed the huge Allied bombardments. At the second or third stage in their defensive tactics the German commanders sent small patrols behind Allied positions forcing the British to defend their flanks.

German infantry attack.

German defenders prepare ambush for British troops. (*Imperial War Museum*)

The PRO/WO208/393 has another German report: 'The British infantryman is distinguished more by physical endurance than by special bravery. The impetuous attack, executed with dash, is foreign to him. He is sensitive to energetic counter-attack.'

Max Hastings, in his superb book *Overlord*, wrote:

> Again and again a single German tank, a handful of infantry with an 88mm gun, a hastily mounted attack, stopped a thoroughly organised Allied advance dead in its tracks. German leadership at corps level and above was often little better than that of the Allies and sometimes markedly worse [perhaps obeying a foolhardy order from their Führer?] But at regimental level and below, it was superb. The German army appeared to have access to a bottomless reservoir of brave, able and quick thinking colonels commanding battlegroups and of NCOs capable of directing the defence of an entire sector of the front.

Every army, since warfare was chronicled in the Old Testament, has had the front line soldiers claiming bitterly that their weaponry, uniforms and rations were inferior to those of the enemy.

However, apart from the Allied artillery, which was highly regarded and feared by the opposition, in every single field the German armoury

British armour and infantry.

was more effective. Take tanks: Tigers and Panthers or Shermans and Cromwells. Take machine guns: Spandaus and Schmeissers or Brens and Stens. A German infantry company *carried* sixteen machine guns compared to the British nine and the American eleven. Take the German panzerfaust (the finest infantry anti-tank weapon of the Second World War) or the Heath Robinson PIAT. Take mortars: the British 2-inch and 3-inch weapons were reliable but slow. The German infantry division usually had twenty 120mm mortars, which could hurl a 35-pound bomb (compared to the famous British 25-pounder gun) a distance of 6,000 yards.

The Allied PBI hated and feared the nebelwerfer, the five-barrelled projector whose bombs were fitted with a brilliantly conceived siren, causing it to wail as it flew through the air, which had an effect on those who heard it, often more penetrating than its explosive power. Nebelwerfers came in three sizes: 150mm (75-pound bomb, 7,300-yard range); 210mm (248-pound bomb, 8,600-yard range); 300mm (277-pound bomb, 5,000-yard range). Each of the five regiments of these – the bulk of them concentrated in the British sector in Normandy – contained sixty or seventy projectors, some of them track mounted.

The large, cumbersome, lethal dual-purpose 88mm AA/anti-tank gun ruled the armoured battlefields and brewed up all Allied tanks in its sights.

German Panzer Mk 5, or Panther.

88mm dual-purpose gun.

Nevertheless Allied rations were regarded as much superior to captured liverwurst and other delicacies. After all 'an army marches on its stomach'!

General Brian Horrocks, the best corps commander in the BLA, despite a lacklustre performance in Operation Market Garden, commanded 30th Corps, which in Operation Veritable in February 1945 mustered 200,000 men-at-arms. He wrote sympathetically about the PBI:

> This battle was a particularly good example The 53rd Welsh Division and, further south, the 51st Highland Division were fighting their way through that sinister black Reichswald Forest. Their forward troops would very often consist of two young men, crouching together in a fox-hole, both of whom had long since come to the conclusion that the glories of war had been much over-written. They were quite alone for they might not be able to see even the other members of their own section and all around them was the menace of hidden mines.
>
> It is this sinister emptiness that depresses them most – no living thing in sight. During training, officers and NCOs had been running round the whole time, but they cannot do it now to anything like the same extent, or they won't live long. Our two young men are almost certainly cold, miserable and hungry, but they are at least reasonably safe as long as they remain in their fox-hole. But they know that soon they will have to emerge into the open to attack. Then the seemingly empty battlefield will erupt into sudden and violent life. When that moment arrives they must force themselves forward with a sickening feeling in the pit of their stomachs, fighting an almost uncontrollable urge to fling themselves down as close to the earth as they can get. Even then they are still alone amidst all the fury: carrying their loneliness with them.

Later on in this book, the chapter on Operation Plunder describes Montgomery's military thrust into the heart of Germany, destined eventually for the Baltic and the liberation of Denmark and Norway. He called it, in his inimitable style, 'cracking around' in Germany. His armoured divisions had perfected the Great Swan through the Low Countries. This was to be the Great German Swan. The author's division was the *schwerpunkt,* the main driving force.

Field Marshal
Bernard Montgomery
with Lt General Horrocks
(30 Corps Commander) right.
Detail of picture on page 7.
(*Imperial War Museum*)

2nd Battalion Grenadier Guards Armoured in the Great Swan. (*Imperial War Museum*)

In a BBC radio broadcast Wynford Vaughan-Thomas described the Great Swan accurately as he watched 11th Armoured Division:

Out on the flanks of our advance go our recce cars [Inns of Court Daimler armoured cars]. Their job is not so much to fight as to find out where the enemy is, and they keep watching the side roads while the regimental group [battle group of, for instance, 3rd Royal Tank Regiment and 4th Battalion King's Shropshire Light Infantry – the author's usual group] is pushing up the main axis [usually called the Centre Line], the road we've chosen for our advance. This group has in the lead perhaps a squadron, perhaps two, of tanks [changed each day to diminish the risk of sudden death] and with a squadron [in fact, infantry platoons] actually riding on the tanks or else carried in their own carriers [Bren or armoured half tracks] are the infantrymen of the armoured division [besides the KSLI, the 1st Herefords or 3rd Monmouths] especially trained to work with the tanks. The tanks of the regimental group HQ [with Lt Colonel Robert Daniel DSO, the author's CO] follow close behind and not so very far behind them come the guns of the group, mobile 25-pounders [Sexton Ram SPs built in Canada, weighing 32 tons, crew of six under a sergeant, maximum speed 36 mph]. There's a gunnery officer [Forward Observation Officer, FOO, usually a Captain RHA] up with the forward elements ready to call up his guns by radio the moment they are needed. [A 'troop' target of four guns; a 'battery' target of eight guns; or, with the colonel's blessing, a 'regimental' target of 24 guns.] So the forward tanks [usually Shermans, occasionally Cromwells] start up the road. When you watch them as I did from a hill overlooking the whole seemingly deserted countryside they seem to probe and stop and hesitate around corners, moving by little fits and starts, for being in the lead tank is one of this war's most uncomfortable jobs. The officer [or sergeant] in that lead tank has got to make absolutely certain, before he moves, that the house up the road, say, is not hiding an anti-tank gun [the deadly 88mm dual purpose gun is a large, free-standing brute that certainly could not be installed inside a house], or that the woods on either side do not contain fox-holes [slit trenches] with the Germans [often young lads of 15 or 16] manning bazookas [cheap, one-shot, hand held, deadly up to 50 yards]. If all is clear then he radios down the line and the column moves on to its next bound. But if the enemy opens up then the flexible structure of the regimental [battle] group allows the commander to decide exactly how he should deal with the opposition. It may be that it's only a small and extremely frightened flak unit [with one or more 88mm guns] firing one round for form's sake before surrendering. The tanks can deal with that roadblock by blasting away at it [the new Comet tank had a 17-pounder gun; the Sherman had a 75mm gun with a lighter shell]. But the defenders may be made of sterner stuff. Then the infantry may have to dismount and work around the road block to outflank it, while tanks and the 25-pounders give them fire support. The regimental group can in some cases call up our fighter-bombers [usually Typhoons. The radio call was 'Limejuice', followed by a specific map reference. A squadron of Typhoons was called a 'Cab Rank'] to hit the strongpoint before they move to the attack. And it may take an hour or so [besides the luxury of the Cab Rank, one alternative was to have available one or more AVRE (Armoured Vehicles Royal Engineers) Churchill tank with a bombard spigot bomb, nicknamed 'the Flying Dustbin', which could destroy most 'fortifications' at up to 200 yards]. Small-scale fighting before the roadblock is smashed and the spearhead probes forward again. So when you watch it on the ground our advance seems no wild race but rather a series of little actions fought along the road almost in fits and starts.

In a curious sort of way the Great Swan was a fairly civilised form of warfare. Casualties on both sides were fairly light. Relatively little damage was done to the villages and small towns along the armoured divisions' centre line.

11th Armoured tanks approaching Antwerp.

However, the more traditional Montgomery set-piece grand slam operation was another matter; a long intense bombardment, initially perhaps by the RAF or USAAF, followed by a huge artillery barrage involving 25-pounder regiments, medium artillery, even heavy artillery and occasionally the Canadian rocket 'flying mattress', all fairly skilfully directed at the enemy defences identified by RAF photography. Then a combination of infantry and armour set off, each offering mutual support; the PBI 'leaning on the barrage', i.e. keeping as close as possible to the shells falling just ahead so that the defenders would keep their heads down. It was all textbook stuff. The only problem was, certainly in a dozen battles in Normandy, that the Germans knew what was coming by observation, listening to radio traffic, and sometimes by Luftwaffe photography. So, in the night they pulled back 90 per cent of their troops to secondary defences perhaps a quarter or half a mile further back, and thus avoided annihilation.

Montgomery, for all sorts of sensible military and humanitarian reasons, wished to retain his finite resources. But inevitably, fighting Hitler's experienced storm troopers, casualties were heavy. The Public Record Office in Kew has an interesting document (PRO WO106/4348) about two British divisions, which identifies statistics of officer and other rank casualties in the north-west European campaign. Predictably 69 per cent occurred during an attack; 23 per cent in defence; and 8 per cent on patrols. Shelling by guns and particularly mortars accounted for 57 per cent, 35 per cent by machine-guns or rifles, and 6 per cent by mines. Infantry platoon leaders and also infantry company commanders each accounted for about 30 per cent, and their commanding officers for a surprising 18 per cent. Since there were infantry battalion commanders in their late twenties, perhaps this is understandable. Again predictably, close-quarter fighting (within a quarter of a mile) accounted for 40 per cent, then 18 per cent at long range, and 13 per cent during the unhappy time forming up for attacks. A British infantry company officer who landed on D-Day, June 1944 had a 90 per cent chance of being a casualty during the 11 months' war; a 20 per cent chance of being killed, and a 70 per cent chance of being wounded. Other ranks had a 14 per cent chance of being killed, and a 62 per cent chance of being wounded. The casualty figures for the whole front line infantry and most of the armoured units during the campaign showed well over 100 per cent of casualties, with reinforcements and 'recycled' casualties coming back for more!

The BLA spent much time and lost many lives trying to capture the key pivotal town of Caen. Eventually High Command asked for the

Allied air forces to 'take out the town', which they did. The vast bulk of the German defenders had left during the night and thousands of French citizens were killed. Unfortunately, this happened very frequently. In the long campaign in Holland, hundreds of Dutch civilians were killed by pre-attack bombardments. In his diary, Lt General Brian Horrocks wrote that during Operation Veritable he felt compelled to order/ask the RAF to bomb Goch and Cleves, key German towns in the Reichwald. In his book *A Full Life*, he admits to feeling guilty that such attractive old towns should be obliterated along with their inhabitants. Later on, the major city of Bremen was 70 per cent destroyed by bombing, but it still needed four British divisions fighting hard for several days to take the place. Hamburg, an even larger city, was destroyed by a firestorm caused by bombing, which killed 30,000 inhabitants. The mayor was allowed to surrender the city to the Desert Rats without a fight.

And so it went on – every city and most large towns in Hitler's Third Reich were reduced to rubble, including of course, Dresden and Berlin.

The ruins of a German city.

LEGEND
Planned Airborne Landings
Ground Attacks
British Divisions. Sept 17th
German   "   "
Front   "   "
Siegfried Line
Rail Supply Line for V2s

ZUIDER ZEE · Zwolle · AMSTERDAM · Apeldoorn · Deventer · Osnabrück · Rheine · MISC GARRISON TROOPS · Utrecht · Neder Rijn · Rotterdam · Arnhem · Münster · Nijmegen · Bocholt · MISC Pz UNITS REFITTING Hamm · R. Waal · Grove · 82ND · Wesel · Uden · R. Maas · FIRST · 101ST · THE · Dortmund · Walcheren FIFTEENTH ARMY · Tilburg · PARA · Essen · RUHR · Breskens · Eindhoven Venlo · Krefeld · ANTWERP · DÜSSELDORF · Wuppertal · 4 Cdn · R. Scheldt 1 Pol · 2 Cdn · ARMY · Roermond · FIRST CANADIAN ARMY · BRUSSELS · SECOND BR. ARMY · XXX CORPS · COLOGNE · SEVENTH · Maastricht · FIRST U.S. ARMY · Aachen · ARMY · Liège · SCALE OF MILES · N

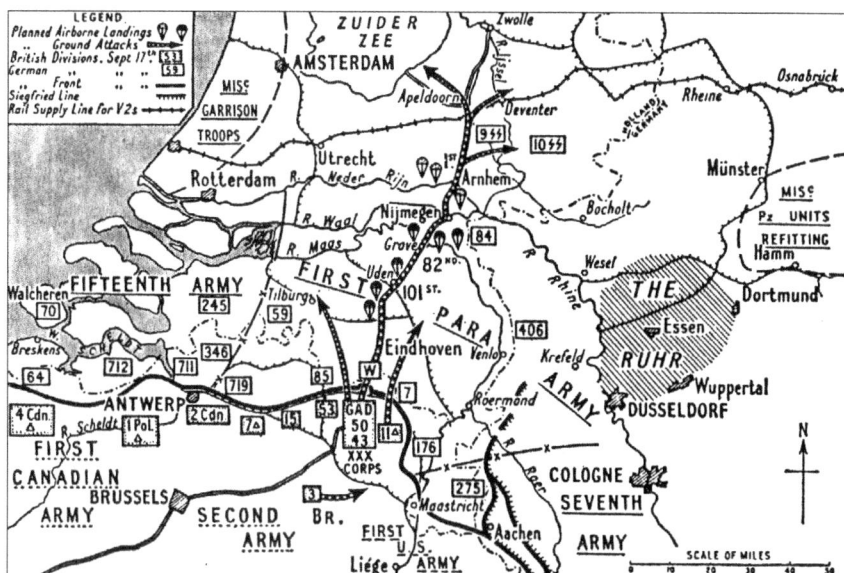

Battleground on the German border. Each square shows a specific German or Allied division.

Towards the end of the war Churchill and most of the Allied generals wanted the joint American and British forces to get to Berlin – symbolically – before the Russians did. Eisenhower consulted with his generals and presumably with SHAEF and asked for their assessment of probable Allied casualties. The consensus was 100,000. Eisenhower arbitrarily decided, without reference to Churchill or Montgomery (but perhaps in consultation with the powerful General Marshal in Washington) that those losses were not acceptable. Ironically, Josef Stalin had no intention of allowing the Western Allies to take Berlin. He did not care if the Red Army lost 100,000, a quarter or half a million casualties – which they did!

Stalin ordered Marshal Zhukov to capture Berlin. It required two gargantuan armies involving 2.5 million troops, 6,250 tanks and 7,500 aircraft to achieve their object. The losses to the Red Army, to the German defenders, and to the inhabitants were unbelievably high.

This author saw many of the shattered villages and small towns in Normandy, battered Bremen (briefly), Hamburg and Kiel. Amid the devastation, the survivors had cleared the streets of rubble, found 'lodgings' in the cellars, and managed to get drinking water and some food in from the countryside. Hans Schmidt-Isserstedt, the gifted conductor of the Hamburg Rundfunk Orchestra, had assembled his orchestra – most

Winter Warfare: white-painted Sherman tanks and marching infantry. (*Imperial War Museum*)

of it anyway – and within a month of the end of war was making superb music for the members of the new British Army of the Rhine!

The moral dilemma of the time had faded. Bomber Harris, RAF supremo, started token bombing of Berlin in 1940. Hitler had upped the stakes with the Baedecker blitzes and when his paratroops, panzers and Stukas invaded Holland, the Luftwaffe obliterated Rotterdam and threatened to do the same to Amsterdam, The Hague and every other city. Bomber Harris was responsible for 1,000 bomber raids. Hermann Göring retaliated with the London blitz, which this author experienced during his school holidays in the West End. The East End and the docks were blown to pieces, as was the open plan Sloane Square tube station, where 250 people were killed. The terrifying aerial battles, tit for tat, went on and on. The end result was the race for nuclear power and the atom bomb; Appendix B details 'Hitler's Nuclear Option'. The Führer believed to the bitter end that his Vengeance weapons, V-1s and V-2s would so cow the British people that Churchill would be forced to sue for peace.

The BLA and those sturdy Cinderella War warriors, Crerar's Canadians, eliminated perhaps 90 per cent of the Vengeance weapon sites. The few surviving sites were in north-west Holland and caused little damage. These weapons were pointless – they could not possibly hit strategic military targets. In effect, they were mechanical suicide bombers targeting everybody and nobody, inflicting awful casual atrocities wherever they landed. The Third Reich will be forever remembered for the Holocaust and for other monstrosities, and the Vengeance weapons should be included in that long list.

CHAPTER 3

Bleak Mid-Winter: Longstop in the Ardennes

Towards the end of Operation Overlord, as the German armies were caught in the Argentan-Falaise corridor, Adolf Hitler started planning for a massive counter-attack, the like of which had not been seen before. The audacious plan was initially called 'Wacht am Rhein' ('watch on the Rhine'), a purely defensive move if the Allies heard about it. His next step was to conjure up the manpower to achieve this secret plan. Hermann Göring proffered his private army of relatively unemployed Luftwaffe ground crew and pilots and summoned a startled Kurt Student, a superb warrior, to head the new 1st Parachute Army, which by 1 December had over 30,000 tough young Nazi fanatics longing to fight and die for their Führer. On 18 October 1944, Hitler announced a gigantic *levée en masse* of the entire German people, ordering every able-bodied man to the defence of the Fatherland.

Anti-aircraft flak units fought as infantry; garrison troops were brought down from Finland, Norway and Denmark; conscripted prisoners of war and the despised 'Ost' troops appeared in their thousands. The Kriegsmarines were first-class opponents and Hitler formed from scratch about thirty brand new Volksgrenadier divisions, which were sent into the line with a few weeks' training. So that was how the Fatherland, in desperate straits by the end of 1944, had over ten million mustered soldiers.

Hitler's proclamation declared that in every district of the greater Reich a German Volkssturm would be set up, comprising every man between the ages of sixteen and sixty capable of bearing arms. These units would be led by the most capable organisers and leaders of 'the well-proven bodies of the party, the SA, the SS, the National Socialist Motorised Corps and the

Hitler Youth'. Although the Volkssturm would not wear uniforms, merely arm-bands with the words 'Deutscher Volkssturm', they were nevertheless soldiers within the meaning of the army code. Service in the Volkssturm was to have priority over duty in any other organisation, and the formation, training and equipment of the Volkssturm was to be the responsibility of the SS Reichsführer Heinrich Himmler. 'The Volkssturm will be sent into the field according to my instructions by the Reichsführer,' ordered Hitler.

'Every mile that our enemies advance into Germany,' said Himmler in a speech accompanying this proclamation, 'will cost them rivers of blood. Every house, every farm, every ditch, every tree and every bush will be defended by men, women and children ... Never and nowhere must or may a man of the Volkssturm capitulate ...'

Brave words, however, were not enough to turn this untrained citizenry into soldiers overnight. It takes training, weapons and time to make a soldier. The Volkssturm could be given very little of each. By the turn of the year its members were still only civilians with arm-bands. The following letter, dated 11 January 1945, written by an old man to his son in the army, might well have been written by one of Britain's Home Guard after Dunkirk:

Every German male up to the age of 60 is liable for service, as is well known. Those who are unfit for military service and those who had been discharged from the Forces are included, if they are fit for office or similar work. ... As we have no equipment or weapons, training presents considerable difficulties, especially during the present weather. Last Sunday week, as we were practicing description of terrain, judging distances and deployment in extended order beside and behind the water-tower (with the high tank), our men suffered very much from the cold. Another point is that the drill is new even for the old soldiers of the last war, whilst many of our men have never been soldiers before. There is need, therefore, for a considerable amount of patience, skill, and above all, time, in order to achieve definite results. Last Sunday we had an army-training demonstration team here, to give our men their first glimpse of the present-day infantry weapons and to explain them in general terms. Much water will have to flow down the Lengenfeld Brook before we receive our own fire-arms and are allowed to take them home.

However, time was the one thing that both the Volkssturm and von Rundstedt did not have.

Wacht am Rhein: by Christmas Day Hitler's three armies had made their furthest advance towards the River Meuse.

Hitler's daring master plan was then called '*Herbstnebel*' or 'Autumn Mist' because it was essential that the operation took place in the worst possible weather to minimise the Allied air forces' 'Jabos'. So D-Day was fixed for 16 December. In 1940, Hitler's armies bypassed the French Maginot defence lines, with their *schwerpunkt* in the Belgian Ardennes. Despite all his generals' advice this is what he did a second time. Two huge Panzer armies, backed by a third Wehrmacht in great secrecy, were assembled and burst upon several scattered US infantry divisions. Ultra at Bletchley Park had picked up a dozen clues, but no senior SHAEF officer pieced them together.

Hitler had assembled this colossal army and its commander was his old, solid, cautious von Rundstedt. No Allied intelligence officer would connect him with an audacious, perhaps foolhardy operation such as Autumn Mist!

In October 1945 Alan Moorehead wrote:

> The Rundstedt plan was one of the most imaginative and daring proposals
> of the war. He [Hitler] resolved to put his fresh and secret Sixth Panzer Army
> straight through the Ardennes Forest and then, having crossed the Meuse, it

would proceed straight to Brussels and Antwerp. The possibilities were endless. To begin with, Eisenhower's forces would be cut in half. Four Allied armies: the Canadian, the British, the 9th US and most of the 1st US, would be bottled up in Holland and Belgium. With Antwerp taken they had no escape route by the sea. Their only hope would be to fight their way south into France, abandoning their vast system of dumps and many, many thousands of guns, tanks, workshops and men. There was even some doubt as to whether the trapped armies could put up much resistance. Once Antwerp fell they had no other supply port. Within a week (so the German planners argued) we might have exhausted our shells. This plan ... was not so fantastic as it sounds.

Moorehead outlined the other key factors. No information available, thus confusion would happen; the German generals knew the Ardennes terrain well; low Allied morale; no point in taking risks if the end of the war appeared to be so soon; the quality and size of Rundstedt's strike forces; the possible renaissance of the Luftwaffe. 'All of these factors together ... and you must concede that Rundstedt had a reasonable hope of so cutting up Eisenhower's armies that they would not recover for many months to come ... Yes anything might happen if the Reich got through intact to the summer of 1945. They might even propose a separate peace to Russia. Hitler might say to Moscow, "Look you see the Allied position in the west is hopeless. Let us come to an agreement and finish the war."'

The Battle of the Bulge became one of the famous (six-week) campaigns of the Second World War. It was America's greatest land battle of all time and the BLA only had a small part to play in it. It was won after attritional fighting in snow, ice, fog, swamps and misery. The GIs, with immense support from the combined air forces – when the skies eventually cleared – won the day.

Field Marshal Montgomery reacted quickly and Lt General Horrocks was ordered to get his 30th Corps in place post-haste to guard all the key towns with bridges across the river Meuse, specifically at Dinant, Namur and Liège. The 6th Airborne Division was back in the UK, but by Boxing Day the 3rd and 5th Parachute Brigades were in place around Rochefort and Grupont.

Both the American and the British press were complaining bitterly that Field Marshal Montgomery had committed no British troops to fight alongside the Americans in the Ardennes. Lt General Horrocks'

Right: Field
Marshal Bernard
Montgomery
inspects 6th
(British) Airborne
Division with
Major General
E. L. Bols and
senior officers,
16 January 1945.
(*Imperial War
Museum*)

Below: Desert Rats
troops produce
home comforts
for a slit trench,
Christmas 1944.

30th Corps, or most of it, had been guarding the vital bridges over the river Meuse since 21 December – a front of 25 miles. For a week 30th Corps acted as an efficient longstop in case the panzers did succeed in reaching the Meuse. With adequate fuel supplies they would have done. Horrocks did not want to cross in force east of the river, which would move them across the American lines of communication. Montgomery urgently required his key 30th Corps for the next phase of 21st Army Group's advance into Germany for Operation Veritable. However, when General Harmon, encouraged by General Hodges, attacked and defeated the German 2nd Panzer Division, Montgomery ordered 30th Corps to cross at Dinant and over a second new bridge at Chanley. The author's armoured brigade of 11th Armoured Division included 3rd Royal Tank Regiment at Dinant, 2nd Fife and Forfar Yeomanry at Namur and 23rd Hussars at Givet. 8th Rifle Brigade actually guarded the three vital bridges.

On Christmas Eve, 3rd RTR were in action against the reconnaissance troops of 2nd Panzer Division and destroyed two Panthers, a Mk IV, a half-track and a scout car. Trucks loaded with ammunition and fuel were satisfactorily blown up. On Christmas Day the veteran 51st Highland Division was detached south of the Meuse near Liège as a reserve to the US 1st Army.

On 5 January, Chester Wilmot sent out his BBC War Report:

British troops are fighting alongside the American First Army in our counter-offensive against the western end and the northern flank of the German salient in the Ardennes. It was British armour and infantry which captured the two villages south of Rochefort on Wednesday and yesterday another British force joined in the attack on the Americans' right. East of Marche they pushed the Germans back nearly a mile with a series of thrusts into the hills along a five to six mile front ... When the Germans broke through in the first weekend of their offensive Field Marshal Montgomery at once brought British divisions south in case the Americans needed them and as the German threat developed, the roads leading back into Belgium were packed with British convoys day and night. It was remarkable how quickly and efficiently we moved thousands of troops with all their guns, tanks, ammunition and engineering supplies. The Belgians lined the roads and cheered the Tommies as the convoys moved towards the new front ... The combined Allied attack brought disaster to the Germans' division thrusting west from Rochefort. Its leading elements were within sight of the Meuse but at Ciney and Celles the British and Americans caught them ... we've counted lying there derelict 81 German tanks, 7 assault guns, 74 other guns and 405 vehicles.

One of Hitler's surprises was the Luftwaffe Operation Bodenplatte, a colossal one-day 'strafe' involving a thousand planes which smashed up dozens of Allied aerodromes and wrote off several hundred planes. It was a thunderclap on New Year's Day 1945. The author was in Brussels when their onslaught savaged three airfields early in the morning, within a mile of him.

On 3–4 January the 53rd Welsh Division, two battalions of the British 6th Airborne Division and the 29th Armoured Brigade were all in action in the combined Allied counter-attacks around Foy-Notre-Dame. Horrocks' attack as right-flank protection to the US 7th Corps was in two prongs. Armour (29th Armoured Brigade early in December had been issued with superb, brand new British-made Comet tanks, equipped with 76mm gun and quite formidable. At panic stations on 18/19 December they had to leave their new toys and retrieve their beat-up Shermans from a tank depot near Brussels) and paratroops

Sherman tanks in the Ardennes.

would attack the German defence line facing north-west along a high ridge eight miles from St Hubert, and also secure, in a mainly infantry attack, crossings over the river Ourthe and cut the vital Rochefort-St Vith road. The villages of Sorinnes and Boiselles were found clear. 61st Reconnaissance and a dashing Belgian SAS unit patrolled eastwards; Hans-sur-Lesse and Tellin were clear but St Hubert and Bure were firmly held. The armour and 7th and 13th Parachute battalions of 5th Parachute Brigade were tasked with the capture of Bure and Wavreille and exploitation to Grupont and Forrières. 3rd Parachute Brigade occupied the area around Rochefort.

Major Jack Watson, CO 'A' Company 13th Parachute Battalion, later wrote:

We reached the start line and looked down on the village which was silent. However, the enemy knew that we were there and were waiting for us. As soon as we broke cover, we came under heavy fire – I looked up and saw the branches of the trees above me being shattered by machine-gun fire and mortar bomb splinters. The enemy had set up sustained fire machine-guns on fixed lines and these pinned us down before we had even left the start line. This was the first time that I had led a company attack and after a few minutes I had lost about a third of my men. We were held up for about a quarter of an hour because of the dead and wounded amongst us but we had to get going.

We were some four hundred yards from the village and so, as quickly as I could, I got a grip of my company and gave the order to advance. Whatever happened, we had to get into the village as quickly as possible. On the way we suffered more casualties, including my batman. One of my men was hit by a bullet which ignited the phosphorous grenades that he was carrying. He was screaming at me to shoot him. He died later.

We reached the village and took the first few houses, in one of which I installed my company headquarters. At this stage I was unaware that B Company was also suffering badly in its attempt to take the high ground, having come under fire from tanks and artillery. Bill Grantham had been killed on the start line, along with one of his platoon commanders. Lt Tim Qinser and Company Sergeant Major Moss, his company second in command and one of the other two platoon commanders, had been wounded. The only surviving platoon commander, Lt Alf Largeren, led the remainder of B Company to their objective. Unfortunately he was killed later in the day whilst clearing a house in which there was an enemy machine-gun position.

Once the company was in the village, it was very difficult to find out exactly what was happening. I got my platoon commanders together to make sure that their platoons were secure and to give them orders for moving forward to start clearing the rest of the village. It was a most peculiar battle because we would be in one house, with my company headquarters and myself on the ground floor, when my radio operator would suddenly tell me that there were enemy upstairs. In other instances, we were upstairs and the enemy were downstairs!

We were suffering heavy casualties as we advanced, clearing each house in turn. Eventually we reached the crossroads in the middle of the village by the old church. I had kept the commanding officer informed as to our progress and he decided to move C Company up to support us. However, by that time the enemy had decided to send in some of their Tiger tanks and they were now firing at us, demolishing some of the houses in the process. I moved from one side of the road to the other, drawing their fire. One of the tanks opened fire on me and the next thing I knew was that the wall behind me was collapsing. At that point one of my PIAT teams came running up and opened fire on the tank, destroying its tracks. They were extremely brave.

The battle went on like this all day. The enemy counter-attacked but we managed to hold them. They forced us back and then we in turn advanced again. It became very difficult to keep the men awake because they were exhausted and had not eaten a hot meal all day. During that night, the fighting continued non-stop with us firing at the Germans and them firing back and shelling us. When we had told Battalion Headquarters that we were up against armour, C Squadron of the Fife and Forfar Yeomanry was sent forward to support us. A troop approached the village first but the leading tank was blown up by a mine. The remaining tanks then tried to approach from another direction and reached the village but one was knocked out. Three tanks stayed with us during that night but by the following morning all of them had been knocked out. The problem was that the Shermans were no match for the Tigers and by the end of the battle 16 of them had been put out of action. We were also reinforced by C Company of the 2nd Battalion of the Oxfordshire and Buckinghamshire Light Infantry, under Major Johnny Granville, because my own company was by then down to one platoon in strength.

During the following day, 5 January, we were subjected to five further counter-attacks which were supported by enemy armour. However, by that time we had a field regiment and a medium battery in support and started giving the enemy a hard time too. They responded to our shelling by trying to blast us out of Bure with their own artillery. Luckily, however, most of my men had experienced heavy shelling in Normandy and thus knew what to expect.

I instructed Major Johnny Granville to move his men forward beyond my own positions in order to find out what was happening. As his company advanced, the enemy counter-attacked again with two Tiger tanks in support. We beat them off and then everything went quiet. At that point I decided that it was now high time that we secured the other half of the village, along with C Company and Major Granville's company, and cleared the Germans out of all the houses. This we did, with much hand-to-hand and close-quarter fighting going on all day.

By about 2100 hours that night we had finally taken the whole village, with my company overcoming the last enemy position. We established ourselves in defensive positions but that same night received orders to withdraw. We later discovered that 7th Parachute Battalion had made an approach from a different direction and, meeting little opposition, had taken Grupont. As a result, we ourselves did not have to go any further. In the very early hours of the following day, 6 January, I assembled my very tired and very wet company and withdrew to Tellin. The battalion had suffered casualties which totalled seven officers and 182 soldiers. Of these, about 68 men had been killed, of whom about half were my company. They were buried in a field in Bure by our padre, the Reverent Whitfield Foy, a few days later.

Lt Edward Harte, 23rd Hussars, wrote:

A Squadron moved out from Beauraing, then to Wellin and Tellin. The awkward top-heavy Shermans skated about on the icy roads like a stampede of drunken elephants. First Sergeant Huthwaite's tank went up on a mine; next Sergeant Roberts' was bazooka'd and he was killed. When we reached Bure four more tank crews were killed. The Germans clung to the houses and ruins, hid in cellars and catacombs, fighting and sniping to the end. There followed an afternoon of very bitter fighting in the village which was in a hollow and the main street was littered with bodies, both Airborne and Germans. Stanley Goss with his troop gave magnificent support but both his support tanks were knocked out by a Tiger.

The 53rd Welsh Division arrived on New Year's Eve. Their 160th Brigade relieved 2nd US Armoured Division in the Marche area and 158th Brigade relieved 2nd US Armoured Division on the river Lesse near Houyet and Ciergnon. Eventually a 13-mile front was held between Houyet and Aye. There was thick snow on the ground and icy, dangerous roads. The

Highland Light Infantry regimental band, about forty strong, came out for a jolly visit over Hogmanay and soon found themselves in the line where they could see *live* Germans within 200 yards! Their two gunner regiments, 83rd Field at Fromville and 133rd at Rochefort, were soon in action. Along the roads lay wrecks of German vehicles of all sorts abandoned, burnt and shattered, but also many American anti-tank guns over-run by the enemy and shot up at close quarters. CSM Cullen, 4th Welch Battalion, noted: 'We were moving across a long viaduct with a concrete balustrade along each side when we saw tanks coming down the road towards us. They were American and in a real state of panic. Crewmen were standing with their heads visible in the turrets and were shouting that the Krauts were behind them. The tanks were moving faster than they should have been on the icy surface.' One Sherman, by mistake, crushed a Welsh rifleman to death, pinning him to the side wall. 'Later we found that the American tanks were supposed to be there to provide us with support in establishing new positions.' The main opposition ahead was the remnant of 116th Panzer Division with six battalions, each reduced to about 300 men backed by some tanks, SPs and mortars. Elements of 9th Panzer Division were in the south-west sector opposite Marche. 53rd Welsh Division then took part in a four-day battle to capture Rendieux-le-Bas, Waharday, Grimbiemont, and the line of the river Hedrée.

There is a detailed account of the very difficult attacks against the 2nd Monmouthshires, 1st East Lancashires, 6th and 7th Royal Welch Fusiliers, 4th Welch, the Oxfordshire and Buckinghamshire Light Infantry and the Manchesters in *Red Crown and Dragon*. There are two interesting accounts, the first by Major A. J. Lewis, 4th Welch, which was tasked with the capture of Authiers de Tailles:

> The rough mountains would have made the going hard in any case, but snow, ice and the cold made the going even worse. There was no cover and the icy, cold wind seemed to whip right through our bodies. The attack was made [supported by 144 RAC tanks] in the face of heavy machine-gun and artillery fire and many were the deeds of outstanding heroism. One private soldier [Private J. A. Strawbridge] after being hit was seen to carry back wounded on five occasions before inevitably being hit again and mortally wounded. A lance corporal with his lower arm blown off continued to lead his section in attack until the objective was gained. How could the enemy withstand such courage! Neither soldier received any award for gallantry.

The Welsh 53rd
Division advance
towards Hotton,
9 January 1945.
(*Imperial War
Museum*)

British Shermans
of 29th Armoured
Brigade (11th
Armoured Division)
advance towards
Hotton, 9 January
1945. (*Imperial War
Museum*)

The second account is in the diary of Lt Colonel Crozier, CO of the 1st
Manchesters:

> Jan 1: 71 Brigade at Marche taking over from the Americans C Company.
> Move with 158 Brigade and take over on the left. Jan 2: All companies had a
> dreadful day trying to get their carriers up, some only making one mile in eight
> hours, roads covered with ice. Temperature dropped. Very cold. Jan 3: Moved
> HQ to Sinsin-Granite, 8 miles NW of Marche. Jan 4: Attack today went well
> in spite of considerable enemy opposition and very bad weather conditions.
> Several enemy counter-attacks including Tiger tanks. Snowing all day, about
> three inches on ground. Going very bad. C Company abandoned their carriers
> and carried their [MMG] guns and ammo up two miles of hillside. Jan 5: 2
> Mons after several attacks failed to take their objective on the left and this pm
> the general [Ross] decided to switch the attack to the right and centre. Nothing
> gained all day and enemy counter-attack this evening against 7 RWF met with
> some success. Heavy snow in the hills and frosts south of Marche and Hotton.
> Jan 6: Last night's counter-attack postponed our attack. Jan 7: Very successful
> battle today, all objectives taken. Division is now on high ground N of Marche-
> Rendieux road. Our infantry casualties fairly heavy.

Indeed, the division suffered well over a thousand battle casualties in
their week of fighting, plus several hundred cases of frostbite. Although
the Airborne Division, fresh out of the UK, arrived with white camouflage
winter clothing, the Welsh PBI were issued with theirs *after* the Ardennes
campaign! Having taken Waharday and cleared the road blocks on the
road to Rendieux-le-Bas, the 53rd Welsh Division went into reserve on
11 January in the Liège area and was replaced by the 51st Highland
Division.

Arriving on 7 January in the Ardennes, the 'Highway Decorators'
were in action at dawn on 9 January. Their first objective was the high
ground west of La Roche, 30 miles east of Dinant and 15 miles north-
west of Bastogne. The Scotsmen had a good supply of 'weasels', little
tracked vehicles that could move across snow without bogging down.
The supporting tanks of 33rd Armoured Brigade had special gripping
snow studs fitted to their tracks. Major Martin Lindsay, acting CO of 1st
Gordons, visited the RWF battalions that were holding the front until
relief on 9 January. They recommended a *constant* supply of dry socks
and hot tea! By nightfall Warizy, Cheoux, Hodister and Lignièrie were

Troops of the 5th Seaforth Highlanders of 51st Division move through Hotton in Ram Kangaroos in January 1945.

51st Highland Division in Hotton on the way to La Roche (*Imperial War Museum*)

captured or occupied. The Derbyshire Yeomanry's armoured cars led the Black Watch into Hampteau and at dawn on the 11th into La Roche. Pte Stan Whitehouse, B Company 1st Black Watch, recalled: 'We occupied La Roche village littered with German corpses from an artillery shelling. In the sub-zero temperature the bodies looked fresh and still alive. We were ordered to deal with SS men still holding out.' In front of the men of B Company, two platoon commanders were shot by snipers. 'Once again we were without an officer. Platoon commanders were snipers' prime targets, a lesson we had learned in Normandy, but a new breed of officers coming through made the enemy sharpshooters' task so much easier by flaunting themselves in the front line with map boards, binoculars and other trappings of rank.' The Highlanders pushed on to Hives, Lavaux, Thimont and Roupage. 152nd Brigade was sent over the hills to occupy Halleux and Vecmont. Divisional HQ moved to Rendeux Haut to control the final phases of the battle. The enemy, from 3rd Panzergrenadier Division and 116th Panzer, were now retreating fast to keep pressure on the Bastogne defenders. But they left small rearguards, snipers, mines and booby traps to delay the Scotsmen. With 32 degrees of frost, Bren guns froze up and armoured car and tank engines had to be run to avoid seizing up. By 15 January, at the end of their battle, 5th Black Watch had lost six killed in action, thirty wounded and forty-two sick and out of action with frostbite and exposure to the intense cold. On 10 January, Genes and Ronchamps were occupied. Major H. Decker, a gunner FOO with 5th Seaforth and a former bookmaker, had laid odds against 'his' Scots capturing Mierchamps. From Ronchamps much of the mile-long road to Mierchamps was heavily mined and there was a blown bridge on the way over the little river Brouze. Up a winding road through thick woods and over a ridge to the village where many panzergrenadiers were lying in wait, odds of 50:1 looked about right. However, Decker with his artillery regiment laid on a thundering barrage and by 1900 hours, to everyone's surprise, 180 half-frozen panzergrenadiers defending Mierchamps were in the bag.

The advance went on. 2nd Seaforth attacked and captured Ronchamps but suffered casualties from heavy shelling. Spandaus, mortars and Panthers caused casualties to the Argylls trying to capture Beaulieu. On 13 January, they not only took Beaulieu but Lavaux and Cens as well. Major Martin Lindsay: '13 January was an unlucky, unpleasant day.' From La Roche, 1st Gordons were ordered to go through Hubermont and then occupy Nisramont. The score of panzergrenadiers in Hubermont put up a fight; and 'friendly fire' from the Black Watch inadvertently wounded Lt David Scott-Moncrieff. Panther tanks in Nisramont knocked out four 'friendly' tanks and SPs; Frank Philips, their gunner FOO was badly wounded. Every Gordon's vehicle was shot at by the hostile Panthers and Major Martin's men suffered twenty casualties. On 14 January, it was the end of the battle for the Highland Division when they linked up with 84th US 1st Infantry Division advancing northwards. Two days later the US 1st Army and the US 3rd Army met at Houffalize, seven miles east of Nisramont. Monty's British 'longstop' troops had suffered 1,400 casualties including 200 killed in action in their three weeks of minor battles closing out the very tip of Wacht am Rhein. They had helped to push back the panzer forces almost 40 miles from Dinant. Ewen Traill, the padre with 1st Gordons, always of his own choice went into battle as a non-combatant with one or other of the rifle companies. He said, 'What a small degree there was between sticking it out and breaking down. When going forward, it only needed one man to shout, "This is murder, I'm getting out" and he would take half a dozen with him.' The margin between success and failure was narrow, with tired troops and war-weary NCOs who did not feel up to taking the responsibilities of officers.

The war correspondent for the *Sunday Times*, R. W. Thompson, noted the beauty of the snow-covered Ardennes hills, but 'with every mile forward, this loveliness becomes a menace and a horror to fight with all the energy each man can muster ... up every hill the troops are manhandling the heavy trucks trying to gain a wheel grip even with chains. Here and there the tracked vehicles slither hopelessly to subside deep into the ditches ... But all the time bulldozers are working, clearing and breaking up the snow and ice to powder and civilians are smashing away with picks and shovels while every man with a spade digs down to the earthy roadside banks beneath the snow to shovel soil for the wheels that must grip ... their ears are blasted constantly by shock as heavy guns roar and splash the white world with bursts of flame.'

The most controversial event that occurred during the Battle of the Bulge was not when General Eisenhower felt compelled to ask Field Marshal Montgomery to take temporary command of two American armies. That decision was logical because the German hammer blow of ten panzer divisions with 2,000 tanks and SP guns had split the American

The harsh winter of 1944–45.

defences down the middle and somebody had to give them sensible orders. Since Monty and Ike had been exchanging 'words' for several months on policies and priorities, it was very much to Eisenhower's credit that he asked Montgomery to take command of General Simpson's US 9th Army and General Hodge's 1st Army on the north-east flank of the battle. This Monty proceeded to do, efficiently and very politely and tactfully so that the two American generals could have no complaint. However, after the battle was over he obviously caused deep umbrage in interviews when discussing '*tidying*' up the defensive lines and writing: 'I am enjoying a very interesting battle, but one ought to burst into tears at the tragedy of the whole thing'. The conference on 7 January was a disaster for public relations.

In any case it ended as a great American victory. However, the history books rarely mention that the ferocious battle lasted for *six weeks*, to the end of January 1945; that General 'Blood and Guts' Patton wrote in his diary on 4 January just after the 6th US Armoured Division was severely mauled, 'We can still lose this war'; that Winston Churchill took such a grave view of the probable outcome that on 6 January he begged Josef Stalin to advance by a week a major winter campaign (which he did) to take the pressure off the Ardennes (Hitler immediately sent reinforcements to the Eastern Front); that Major General 'Slim Jim' Gavin, the highly experienced GOC of the 82nd US Airborne Division wrote: 'We are training our [airborne] men to drive tanks and tank destroyers, since our armoured supporting people frequently abandon their vehicles when threatened in an attack. If our infantry would fight this war would be over by now ... American infantry just simply will not fight. No one wants to get killed. Our artillery are wonderful and our air corps not bad, but the regular infantry – terrible.'

Nevertheless in the wicked winter conditions of deep snow, icy winds, icy roads and often thick fog the American GIs rallied. The skies brightened and the Allied fighter-bombers ruled the battlefields and very, very slowly the three great German armies, by the end of January and after six weeks of ferocious fighting, were back where they started.

Wacht am Rhein had not succeeded. The losses on both sides were hideous – between 80,000 and 90,000 each.

The Ardennes struggle had ensured that General Eisenhower would take even fewer risks. His nerve had been badly shaken. It took seven weeks for Eisenhower's armies to recover their balance after the shock of the Ardennes Battle of the Bulge. The Führer's most formidable forces, 5th and 6th Panzer Armies, were mauled and unable to oppose Marshal Zhukov and his Russian forces in the Vistula offensive early in 1945. General Horrocks, in his book *A Full Life* wrote:

The Germans had failed primarily because they had underestimated the fighting qualities of the American front-line troops. The American divisions had weathered this unexpected storm most creditably, but their losses had not been light ... The German losses in men and material were, however, much more serious because, unlike the American, they could not be replaced. In the words of Manteuffel himself: 'The cost was so great that the offensive failed to show a profit. The last German reserves had suffered such losses that they were no longer capable of affecting the situation either on the western or eastern fronts.'

The effect on the morale of the German troops was disastrous. After the Ardennes all hope of winning the war had gone. Disillusionment and bitterness now began to creep in.

CHAPTER 4

Hobart's Terrible 'Funnies'

Percy Cleghorn Stanley Hobart, known to his family as Patrick and to the British Army as Hobo, was a remarkable man. He had many claims to fame. He was the only man in the world to have created and formed three magnificent fighting formations. In late 1938, he raised more or less from scratch Britain's first armoured division. The 7th Armoured fought in all the turbulent North African campaigns, winning three Victoria Crosses in the process. However, Hobo, who was the most experienced tank man in the British Army, was booted out of the army in late 1939 by a 'Junta' of cavalry generals and joined Churchill's Home Guard as a corporal. Rescued by Churchill personally, he was then entrusted with the formation and training of the 11th Armoured Division, to which the author belonged – this time mainly in Yorkshire. The Hobart family had a bull as part of their escutcheon and a rampant black bull with red hooves, horns, nose etc., duly appeared. Destined to sail with the division to North Africa, Hobart failed his medical test. Churchill then secured for his protégé the responsibility for creating a brand new concept – a military zoo of either very dangerous or very useful animals!

The 79th Armoured Division – the Germans were not fooled by the British Army's apparent luxury of quite so many divisions – became by far the largest armoured division in the world, with 21,000 soldier-engineers. It never fought in any battle as a division, not even as brigades, but usually as regiments or squadrons. It also had the insignia, in a triangle, of a thoroughly bad-tempered bull, black, white and dark orange. In fact Hobart's brief was, 'Your formation will be trained primarily in their special roles and secondly as normal tank formations'. Every 'Funny', as they were christened by the army, had to have an offensive characteristic.

Hobart's crest for his 79th Armoured division.

Crocodile flame-thrower with trailer, in action north of Schilberg during Operation Blackcock. (*Imperial War Museum*)

The most fearful zoo animal was the Crocodile flame-thrower, which in 1940 was assembled by the Lagonda car company with technical assistance from the Anglo-Indian Oil Company. A variety of prototypes were created, including the RNR Cockatrice, which was designed for airfield defence against enemy air force troops. In 1941, the 7th Buffs were converted from infantry into the 141st Regiment Armoured Corps (RAC) and with 7th RTR and 9th RTR formed 31st Armoured Brigade. The British-made Churchill infantry-support tank Mk VII had a flame nozzle, which replaced the hull-mounted machine-gun, so its main 6-pdr gun could still be used. A trailer was towed, which carried 400 gallons of diesel oil mixed with tar. Fuel consumption was 4 gallons per second. The duration of a flame shot at a target was only one or one and a half seconds. In the armoured trailer were also bottles of nitrogen for pressurising the fuel, which passed through a pipe fitted under the hull of the tank. The pipe was protected by an armoured shield. The most effective flame range was 90 yards and the maximum 120 yards. It took a minimum of one and a half hours to refuel and usually 15 minutes to 'pressure up', but less in an emergency. The equipment (trailer, nozzle, piping and controls) was manufactured in kit form for the Royal Electrical and Mechanical Engineers (REME) to convert or replace in the field. Eight hundred kits were manufactured. By the end of the war, three regiments of Crocodiles were in action: the Buffs, 1st Fife and Forfarshire Yeomanry and the 7th RTR. Each regiment had sixty Crocodiles and the Buffs only had theirs ready a few days before D-Day in June 1944. The vast majority of infantry commanders were unfamiliar with the flame-throwing concept. At close range they were lethal on enemy pill-boxes, strongpoints, occupied houses, Spandau positions and trenches. They could burn most woods, except those with thick foliage, and the flame could burn on water and across most canals.

Lt Andrew Wilson of the Buffs, author of *Flame Thrower*, describes how he first encountered a Crocodile demonstrating its ferocity:

The crew manipulated a system of valves and gauges. Instead of the co-driver's MG there was now an ugly little nozzle with two metal tongues above it like the points of a sparking plug. There was a continuous hissing and ticking from the trailer. A little burst of fire, like a struck match above the nozzle, tested the spark and the tank began to move forward. It went towards the first target, a concrete pill-box. Suddenly there was rushing in the air, a vicious hiss. From the front of the tank a burning yellow rod shot out. Out and out it went, up and up with the noise like the slapping of a thick leather strap. The rod curved and started to drop, throwing off burning particles. It struck the concrete with a violent smack. A dozen yellow fingers leapt out from the point of impact searching for cracks and apertures. All at once the pill box was engulfed in fire – belching, twisting, red-roaring fire. And clouds of queer-smelling grey-black smoke ... the rod went clean through an embrasure, smacking, belching, roaring.

Quite terrifying! Their weakness was the necessity for close-quarter action and the infantry being supported had to assault the enemy position as quickly as possible after flaming to obtain the maximum demoralising effect. The battle of 's-Hertogenbosch, Operation Guy Fawkes, at the end of October 1944, was a great success for the 53rd Welsh Division and the Desert Rats. Major R. N. Deane, A Company 2nd Monmouths, wrote: 'The thing which most added a new element to the atmosphere of Modern War was Flame. The place at which flame-throwers were principally used during this successful three-day set-piece attack was in the advance on the dykeland village of Bruggen ... there I first saw Modern War closely resemble what I imagine Medieval Hell was thought to be like.'

The Americans were shown most of Hobart's 'Funnies' before Operation Overlord and were reluctant to accept them. However, in the difficult siege of the great port of Brest in Brittany the defenders in the key citadel showed absolutely no signs of surrendering. So General Omar Bradley asked for some Crocodiles to help reduce Fort Montbarey and its 250-strong diehard garrison. A deep moat, an anti-tank ditch plus casemented fortress walls, many guns, three lines of defences and a minefield full of large naval shells combined to make an assault very difficult. Major Nigel Ryle's B Squadron of the Buffs was loaned to General D. H. Gerhardt's investing forces. The Buffs travelled on huge Diamond T tank transporters. The amazing battle that followed is related in *Churchill's Secret Weapons*. Bravery and dash were successful and on 18 September the garrison of 30,000 surrendered; praise was heaped on the Buffs. Generals Simpson, Sands and Gerhardt all wrote commendations; they were featured on the BBC and in *Life* magazine; and they were showered with thirteen Silver and Bronze Stars. The Buffs spent a few days unofficial UK leave and so

AVRE 'bombard' tank with menacing Spigot Mortar Bombard, known as the 'flying dustbin' petard. Ideal for destroying enemy strong points at 100 yards' range. (*Imperial War Museum*)

Sherman flail tank (Crab) crosses the river Orne during operations in July 1944; their mine sweeping saved many lives. (*Imperial War Museum*)

noticeably enjoyed themselves that they were known, forever, as 'the Playboys'.

The second most ferocious 'Funny' was the AVRE Churchill Mk III or IV tank, with a specially designed 290mm calibre Petard spigot mortar with a fire rate of two or three of these huge mortar bombs per minute. The 40-pound projectile housed a 26-pound hollow charge bomb nicknamed 'flying dustbin'. It had an effective range of 80 yards and the spigot was fixed to the 6-pdr gun mantlet. The spare Besa machine-gun was removed to permit stowage of engineering equipment and spare flying dustbins. The hinged hatches over the co-driver's position were removed and the aperture sealed with a steel plate in which a sliding hatch was fitted through which the petard was loaded. The AVRE Churchill's weight was about 40 tons and the crew consisted of the commander, driver, demolition engineer, wireless operator, mortar gunner and the co-driver, who was also the mortar loader.

The 1st Assault Brigade RE consisted of 5th, 6th and 42nd Assault Regiments, usually farmed out a troop at a time. The Churchill tank was made to be extremely versatile and alternative 'Funny' specialities were as follows.

On the back could be carried a huge brushwork paling fascine of which Julius Caesar would have been proud. It was ideal for filling deep ditches and bomb craters, quickly, efficiently and cheaply. The fascines were about 8 feet in diameter and 12–14 feet long. Alternatively a bobbin log 'carpet' could be carried to lay over treacherous muddy ground.

Or a small box iron girder bridge could be carried for laying across streams and minor rivers that did not require a large Bailey (meccano-type) bridge that would take sappers many hours to assemble. It was known as the SBG Assault Bridge.

The Skid Bailey Bridge was pulled or towed to span a 60-foot wide river – it was used in Operation Blackcock, the capture of Bremen and other actions.

A mobile Bailey 'Scissors' bridge was pushed forward by the Churchill tank, disengaged and then mounted by the AVRE to drop the second half in place.

An armoured sledge was towed behind an AVRE, bringing up to the front line heavy engineering gadgets, including explosives.

The Germans were extremely skilled in laying giant defensive minefields, usually of round flat Teller mines, which could obliterate a

Infantry into battle in Kangaroos on the way to 's-Hertogenbosch. (*Imperial War Museum*)

man and knock out any armoured fighting vehicles (AFVs). The author's FOO Bren gun carrier went up on a double Teller mine in the Dutch Peel country, which destroyed the carrier and the unfortunate driver.

The answer was the Sherman Mk V 'Crab' flail tank, which must have saved the lives of hundreds of sappers, pioneers and PBI. General Irwin Rommel had four *million* mines laid along the French coastline defences!

The first flail was a Matilda tank of 42nd RTR, known as a Scorpion and it was deployed in the desert battles including at Alamein. Major W. R. Birt, 22nd Dragoons, who had sorrowfully left their cavalry horses behind and regretfully exchanged their normal Sherman tank assault role for flails, wrote:

> The principal problems in the use of the flail: (1) to maintain a straight and accurate path to the objective when the tank driver was completely blinded by the swirling chains; (2) to direct three tanks together through a minefield clearing a path of mines wide enough for the follow-up infantry and tanks; (3) to protect the flailing tanks during 15–20 minutes as they crawled [at 1½ mph]

A flail tank in burning Arnhem. (*Imperial War Museum*)

their slow, undeviating way in the face of enemy gunners where most of their
own gunners were blind.

The loud jangling noise, the thick dust storm, the exploding mines were
nevertheless an awe-inspiring frightening sight to the defenders.

Major General Hobart realised that a major task was that of
transporting armoured and fighting carriers of the PBI towards and
close to the *schwerpunkt* of their assault. So a magnificent zoo soon
appeared: Buffaloes with a Mk II or III Churchill stripped of its turret,
cupola and main gun, which could carry thirty fully equipped infantry,
or a Bren, or Universal carrier and crew of a Weasel or Jeep, or a 6-pdr
anti-tank gun, or a 25-pdr gun and crew, or an Airborne bulldozer or 4
tons of cargo. They had three or four 30-inch machine-guns or a 20mm
Polsten cannon. They had a speed of 7 knots on or in water hazards, or
25 mph on land. There were 600 Buffaloes in Hobart's zoo and they
were very popular with the PBI. They were noisy so movements were

usually made at night. Terrapins were eight-wheeled trucks that could carry 8–10 infantry.

The clever Canadians had built Sherman-style AFVs, which included the Sexton SP 25-pdrs (in the author's regiment) and the turretless Rams, used to carry infantry in Operation Totalise. So Hobart introduced the Kangaroos to 49th RTR in December 1944 and quite quickly 250 vehicles were constructed. They could take a section of eight armed PBI fairly safely at speed into battle. Indeed, every infantry formation in the BLA was hooked on Kangaroos, or if 'swimming' was required, Buffaloes.

The German defenders everywhere were dismayed by the sudden arrival of the British *Schwim-Panzers*, Sherman gun tanks carefully fitted with dual-duplex (DD) steerable propellers. They travelled (swam) at 4 knots and had a high waterproof canvas screen to keep water out. The defenders of the city of Bremen were noticeably disheartened when a DD Sherman attack was launched on them. Once General Montgomery saw the DD tanks in their trials at Linney Head in Pembrokeshire, he decided they would lead the Allied forces ashore on D-Day. In the water they looked like harmless rubber boats. Of the 122 DD tanks launched from their LCTs, 83 landed ashore safely and went into action. The Americans, General Eisenhower in particular, liked the idea, but launched their fleet not 4,000 yards out, but 6,000 yards out in a rough sea, which was asking for trouble. Of the thirty-nine DD tanks that were swamped, the majority were from 749th US Tank Battalion.

Hobart's fecund mind also produced the Bullshorn, Conger, General Wade, Goat, Great Eastern, Lobster, Porpoise, Rhino, Rodent, Snake, Tapeworm, Weasel (an amphibious Jeep) and Wasp (a small flame-thrower on a Universal carrier).

In December 1944, Hobart set up 557th Assault Training Regiment at Gheel in Belgium. It became the Assault Training and Experimental Establishment RE. New problems were the wooden German Schu mines; more sophisticated tank bridges; improved multi-barrelled smoke shell dischargers (vital in operations Veritable and Varsity); indestructible Canadian rollers pushed in front of tanks to clear mines; tapeworm anti-mine devices; a Crocodile to scorch and burn *mined* areas; a Weasel mine-clearer for use against Schu mines; ARKS; log and other carpets; and a tankdozer (a bulldozer mounted on a tank).

Major General Hobart was one of the heroes of the Second World War. In the First World War, he won the DSO and the OBE, was mentioned

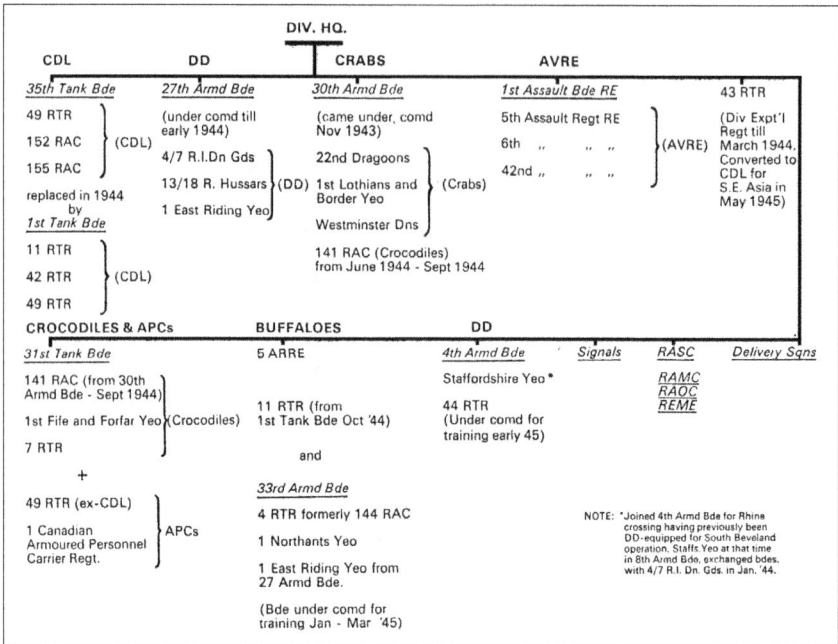

	DIV. HQ.			
CDL	**DD**	**CRABS**	**AVRE**	
35th Tank Bde	*27th Armd Bde*	*30th Armd Bde*	*1st Assault Bde RE*	*43 RTR*
49 RTR	(under comd till early 1944)	(came under, comd Nov 1943)	5th Assault Regt RE	(Div Expt'l Regt till March 1944,
152 RAC } (CDL)	4/7 R.I.Dn Gds }	22nd Dragoons	6th „ „ „ } (AVRE)	Converted to CDL for
155 RAC	13/18 R. Hussars }(DD)	1st Lothians and } (Crabs)	42nd „ „ „	S.E. Asia in May 1945)
replaced in 1944 by *1st Tank Bde*	1 East Riding Yeo	Border Yeo Westminster Dns		
11 RTR		141 RAC (Crocodiles)		
42 RTR } (CDL)		from June 1944 - Sept 1944		
49 RTR				

CROCODILES & APCs	**BUFFALOES**	**DD**			
31st Tank Bde	5 ARRE	*4th Armd Bde*	*Signals*	*RASC*	*Delivery Sqns*
141 RAC (from 30th Armd Bde - Sept 1944) }		Staffordshire Yeo *		*RAMC* *RAOC* *REME*	
1st Fife and Forfar Yeo }(Crocodiles)	11 RTR (from 1st Tank Bde Oct '44)	44 RTR (Under comd for training early 45)			
7 RTR }	and				
+					
49 RTR (ex-CDL) }	*33rd Armd Bde*				
1 Canadian Armoured Personnel Carrier Regt. } APCs	4 RTR formerly 144 RAC				
	1 Northants Yeo				
	1 East Riding Yeo from 27 Armd Bde.				
	(Bde under comd for training Jan - Mar '45)				

NOTE: *Joined 4th Armd Bde for Rhine crossing having previously been DD-equipped for South Beveland operation. Staffs.Yeo at that time in 8th Armd Bde, exchanged bdes. with 4/7 R.I. Dn. Gds. in Jan. '44.

Order of Battle – 79th Armoured Division.

six times in despatches, was wounded, captured by the Turks, rescued by a Rolls Royce armoured car unit, and later became *the* tank expert in the British Army. Even Brigadier Michael Carver, the thrusting armoured commander *par excellence* in the Second World War, wrote of Hobo, 'we exercised under the eagle eye of the fierce brigade commander, the great 'Hobo', Percy Hobart. He was a merciless trainer who drove us all hard and overlooked no detail, his intensity matched by his keen interest in all ranks under his command. He was universally respected, admired and served with enthusiasm.' Carver thought Hobo was also 'a bully'!

CHAPTER 5

Bleak Mid-Winter:
Operation Blackcock

Operation Shears had been planned to take place in December, but the Ardennes Wacht am Rhein caused delay, so it became Operation Blackcock and was duly launched by General Ritchie's 12th Corps in mid-January. Two German divisions, 176th and 183rd Wehrmacht, with much artillery, SP assault guns and paratroop units in support, held a substantial triangle on the Belgian-German borders. They held a line 12 miles north-south between Sittard-Maeseyck-Heinsberg-St Odilienberg-Roermond. On the west side ran the river Maas and Juliana Canal and on the east the river Roer. However, running parallel to the start line was a long, winding river 20 feet across called the Saeffeln and Vloed Beeks, which would cause problems.

General Ritchie's plan for Blackcock was in three phases. The Desert Rats (7th Armoured) on the left flank would strike north-east from Bakenhoven towards Roermond, via Susteren, Echt, Montfort and Linne. Parallel on their right would be 8th Armoured Brigade (Red Fox) and 155th Brigade of 52nd Lowland Division. From Susteren they would sweep east to Heide and the wooded, swampy area known as the Echterbosch. The third phase on the right/east flank was for the rest of 52nd Lowland and 43rd Wessex Wyverns to clear a deep triangle into Germany with Breberen, Scherwalden Rath, inside the Siegfried Line and just north of the key town of Heinsberg, a mile from the river Roer.

The German defenders had had plenty of time to lay huge minefields. In reply Major General Hobart's great team of 'Funnies' would provide Sherman flail tanks (Crabs) to deal with the mines; Crocodile flame-throwers to deal with pill-boxes and defended buildings; and scissor bridges mounted on Churchill tanks to cross streams, small

75

rivers and major bomb sites. If the visibility was reasonable, which it was not, the RAF would supply Typhoon fighter bomber support. Lt Colonel Dallmeyer, CO of the Lothians and Borders (flails) was the 'Funny' adviser to the Desert Rats and the 52nd Lowland. He allocated 'horses for courses': for instance the Buffs Crocodile flame-throwers, A Squadron to the Desert Rats and B Squadron to the Lowlanders; and his own Crabs, A Squadron to 8th Armoured Brigade, B Squadron to the Queens infantry brigade (of 7th Armoured), and C Squadron to 52nd Lowland. Captain Harry Bailey was a tough troop commander who had flamed his way from the Normandy beaches and possessed a sense of humour in his macabre expertise. As described by him the 141st (RAC), the Buffs, Crocodiles had had their Churchill flame-throwers 'tricked out in their cute snow camouflage whitewash on the hull and little frocks from parachutes draped round their darling little turrets. Now they looked chic and *distingué*.'

General Ritchie, like all of Montgomery's corps commanders, always had access to considerable artillery support. For Blackcock he had eight field 25-pdr and six medium and heavy regiments on call.

The Desert Rats Queen's Brigade soon reached Susteren after a scissors bridge over the Vloed Beek at Bakenhoven had been put in place by the

Desert Rats AFVs are painted white for snow warfare before Operation Blackcock.

Churchill AVREs. Thick fog caused confusion to friend and foe as both sides wore white snow smocks. Major John Evans with 1/5th Queens infantry won the DSO in the two-day fight for Susteren, which was depicted by the war artist Bryan de Guingeau in the *London Illustrated News* of 17 February. Their attack from the west took place in pitch darkness and in thick mud. Evans wrote:

> We heard a guttural shout and a single shot was fired. We then adopted our usual tactic of rushing in screaming and shouting, firing from the hip. We ran through the forward part of the town taking about 37 prisoners and not suffering a single casualty … At dawn we discovered three enemy tanks in the town. There were no 6-pdr A/Tank guns. Our own tanks had failed to reach us because of the dykes, drainage ditches and soft ground. There followed some desperate fighting with the tanks demolishing corner street properties on top of our men. Corporal Dolly knocked the track off one tank with a Piat for which he later received the Military Medal. I fired my captured Schmeisser – two short bursts at two tank commanders in their open turrets. It was obvious we were in for a long day and would suffer many casualties.

The acting CO, Major Jack Nangle, arranged that the corps artillery would shell the village to try to knock out the tanks. During the shelling and counter-shelling, Evans was badly wounded and Lt Stone was killed. It was a shambles, but the company held its position in spite of thirty-nine casualties, including all the officers, and twenty-nine Queens were taken prisoner. Against that, forty enemy prisoners were taken and at least that amount killed. Fewer than forty men survived that day. Captain John Franklyn and all the platoon commanders were killed. John Evans was the only officer to survive, although he had two sets of wounds and eventually his left arm was amputated. B Squadron 1st RTR finally managed to cross the Beek and come to the assistance of the survivors of B Company; in the mopping up next day they took another seventy-seven prisoners. Peter Roach describes 1st RTR's start to Operation Blackcock:

> We moved off [from Stadtbroek near Sittard on the 16th] on ice-covered roads and in a thick fog which held up the attack and gave us time to whitewash our tanks as some sort of camouflage. With the daylight [on the 17th] came a thaw which though slight turned the tracks to quagmire. Slowly we made our way

forward but the main advance down the centre got bogged down. All through the late afternoon we sat near a stream which the engineers and Pioneer Corps were trying to bridge under constant mortar and shell fire.

In the Susteren battle, 1st RTR lost seven tanks to bazooka teams and anti-tank guns.

The 11th Hussars had put a patrol into Oud Roosteren, which was later captured by 6th KOSB under command from 52nd Lowland Division. Lt Alan Parks, with C Squadron 1st RTR, wrote: 'My orders were to take my troop and capture the small village of Heide, just NE of Susteren, across a railway line as A and B Squadrons were moving north to Echt and Schilberg.' Later C Squadron 1st RTR with the 2nd Devons and flame-throwing Crocodile Churchill tanks, pushed north to capture Ophoven and the western sector of Echt, taking 100 prisoners. The historian of the Devons described the operation:

> In thaw and thick fog at 1630hrs on Jan 17th, the dyke bridges came under heavy shellfire which delayed the tanks. When they got across our carriers had to be hitched to them. In the snow, slush, sniping, across minefields and despite MG and 75mm fire the two columns moved off. Each column consisted of a mobile screen of two troops of 1 RTR tanks, a section of Bren gun carriers, a section of assault Pioneers, then two more troops of tanks, our infantry company in [armoured] Kangaroos, a section of Norfolk Yeomanry SP A/Tk guns, a section of 3-in mortars, a 6-pdr A/Tk section, a RE recce party and the RHA FOO. Major Overton took D Company on the right towards Schilberg. In the villages Nazi slogans were painted on the walls, 'We will never capitulate.'

'Snatch' Boardman wrote in *Tracks in Europe*: '1RTR pushed into Echt and soon met heavy resistance from well-placed A/Tk guns and bazooka men. They lost a number of tanks. The Skins were to pass through Echt and supported by 1/5 Queens, to capture and hold Montfort.' The Germans had several SPs and anti-tank guns in Echt and Schilberg, a mile to the east. By 18 January the 2nd Devons and 1st RTR had taken Echt, Hingen and Schilberg, but three miles north-east St Joost was well defended. Captain Bill Bellamy, 8th Hussars, noted:

> Meanwhile the battle in St Joost itself became fiercer and fiercer. It was a desperate struggle between first-class British troops and one of the toughest

18 January, 1 RTR
and 2nd Devons
capture Echt.

of the German parachute regiments – Hübners. Early on the morning of the
21st another attack was mounted on St Joost. Once again B Squadron under
Wingate Charlton fought with great gallantry in support of the infantry. Thick
fog reduced visibility to less than 100 yards. Richard Anstey and Douglas
Ramf were ordered to take their two troops, retrace the route of the previous
day and find a way round St Joost to outflank it from the east. I was sent to
the bridge in Hingen to act as a forward wireless link if needed. The tanks
clattered off into the dense fog and all went well for some minutes. Then I
heard Richard shout over the wireless 'Two SPs to our front – engaging', and
at the same time I heard the crack of three shots from tank guns, followed by a
series of shots, then silence. [Two of Richard's and one of Douglas's tanks had
been knocked out by three Mark IV 75mm SP guns.] I had an extraordinary
view of St Joost and could see and hear the noise of the battle. It sounded and
looked to be terrifying, flames, smoke, continuous machine-gun and rifle fire,
the crack of tank guns and the whistle and crump of artillery shells.

The three lost tanks were recovered the next day, as well as a Sherman
Firefly from the cellar of a farmhouse!

By nightfall the DLI's fourth attack – D Company plus Crocodiles, 8th
Hussars tanks and A Company 1st RB – had taken the gutted little town.

Hübner later admitted that one of his companies had been destroyed, another almost so.

Although sixty prisoners and three SPs were captured, the losses to the Durhams and RB had been heavy. Most of the several hundred German troops killed were inside the houses or down in their cellars. It was a savage and bitter battle in which Lt Colonel C. A. Holliman DSO, CO 5th RTR and a Desert War veteran, was killed. The Durhams licked their wounds and stayed around Posterholt on the river Roer, temporarily under the command of US 16th Corps. In deep mud they were bombarded by rocket shells which left craters 10 feet wide and 5 feet deep. There they stayed until 21 January, when they retired to Weert in Holland for their first rest since landing in Normandy, in the previous June.

Operation Blackcock still had some way to go, as there were four more defended villages north-east of Echt/Schilberg that had to be cleared: Montfort, nearby Aandenberg, Paarlo and St Odilienberg.

The Skins were ordered to pass through Echt and, supported by 1/5th Queens, to capture and hold Montfort, which had been heavily bombed by the RAF. But C Company 1st RB, commanded by Lt Dawson Bates, reached Aandenberg just north of Montfort, where they spent a hectic 24 hours until relieved first by the Queens, then by the Devons. Corporal 'Snatch' Boardman of the Skins wrote of this action: 'Many bridges were found to have been destroyed but since they spanned narrow gaps they were replaced by scissor bridges though this slowed the pace. The Recce squadron was ordered to relieve 8th Hussars at a stream half a mile from Montfort.'

It was late afternoon on 22 January when the bridge was finally in position and 4th Troop B Squadron crossed. Supported by C Company 1st RB, they moved forward to capture Aandenberg. Montfort presented an eerie sight after the heavy RAF bombardment, with the light from the burning houses and the reflection off low cloud of the searchlights, nicknamed 'Monty's Moonlight'. The troop edged forward and into a desperate battle with the German paratroopers, with the crackle of small arms fire, shouting and the crash of mortar bombs. At 2200hrs the 1/5th Queens arrived on C Squadron tanks and joined the all-night battle.

In the morning, on 23 January, 5th RTR and 2nd Devons finally cleared Montfort, arriving from the wooded western sector, with the Skins, 1/5th Queens and 1st RB moving down from the northern suburbs. Under the ruins lay 270 Dutch civilians killed by the bombing. It had been a two-

21 January 1945. Troops on the right flank of the attack moving north-eastward through Saeffelen and the German village of Hongen.

day battle to clear Aandenberg and Montfort. Even the wood clearing on the 25 January cost 1/5th Queens a further sixteen casualties.

The Skins spent 24–26 January mopping up around Montfort, which was occupied by 1st RB. On the 26th, the final advance to clear the Maas-Roer 'Sittard' triangle continued. To the right, with the objective of Posterhout, a village five miles east, went 1st RTR and 1/5th Queens. In the centre, heading north to take St Odilienberg, were the Skins and 2nd Devons. On the left (west) 8th Hussars supported 1st Commando Brigade to take Linne. 1st Commando Brigade, with 5th RTR, carried out some combined patrols along the Maas near Linne, Heide and 'Bell' island from 25 to 28 January, in which Lance Corporal Harden RAMC won the VC.

Norman Smith, 5th RTR, returning from UK leave noted: 'The Big Freeze came to an end in the first week of February. The Commandos did a good mopping up job and the Maas was crossed by infantry in little boats (some men, incredibly, swimming in the icy water to get the wounded back). A big barrage was put down [on 24 January] while they got the interlinking Bailey bridges across the river for our tanks to go over.' This was in the operation against 'Bell' island on the far bank of the Maas.

81

Lt Hugh Craig-Harvey of the Skins related his experience during the battle for St Odilienberg:

> A set-piece attack by two troops of C Squadron. I commanded First troop and Henry Woods Fourth Troop plus two companies of the Devons, Flail and Crocodile flame-throwers from the Lothians and Border Horse. Since the Americans were expecting to take over the area the following day – and since this was the only serious battle being fought that day (the 26th) a great number of senior British *and American* officers came to watch ... We had been issued with aerial photographs. Why I was not sure, because the village only had two streets and was bounded on the east side by the river Roer.

Craig-Harvey was blown up four times in the five-hour battle. Teller and the wooden Schu mines were everywhere, often under snow. And enemy mortars and machine-guns across the river Roer, which flowed through St Odilienberg, made life unpleasant. Meanwhile, three miles south 1/5th Queens and 1st RTR moved on briskly, and on the same day Posterhout had been secured. By 29 January, the Queens had occupied Paarlo and Holst, south of Posterhout. The Queens were involved in a sharp action at Paarlo. When on the night of the 29th/30th A Company was sharply counter-attacked by fifty enemy in assault boats backed by shelling and mortaring, Derrick Watson wrote: 'The ground was frozen hard so we abandoned the idea of digging in, so Lt Max Baker, a Canloan officer with 8 platoon, occupied a large 4-storey house in the main street overlooking the river.' And Verdon Besley recalled: 'Late in the evening there was a sudden barrage and then a terrific bang upstairs and two men were brought down blinded by a bazooka. Corporal Dennis took them to the cellar for safety.' Verdon was then carrying out a spirited battle from the top of the stairs dodging bullets and grenades, retaliating with both while Dennis did the same from the cellar. Rescue was at hand by a Czech officer known better as Robert Maxwell, who with Derrick Watson and Corporal John Melmoth relieved the beleaguered garrison. The RHA and 1st RTR tanks in support gave the retreating Germans a noisy send-off. The Queens received seven casualties, took ten prisoners and caused many 'hostile' casualties. In March, Monty presented MCs to the two Maxies (Baker and Maxwell) for their part in the Paarlo fracas. On the next day, 31 January, the remaining bridge over the Roer at Vlodrop, south of Roermond, was blown up and Operation Blackcock came to an end – fifteen days and nights of constant fighting against a determined foe, with horrible weather and minefields everywhere.

The Desert Rats, after their sustained brilliance in North Africa, had a difficult time in Normandy and their success in Operation Blackcock helped restore their pride – and fame!

Major General Hakewill Smith's 52nd Lowland Division had arrived late in the day for Operation Eclipse. They originally had been trained for mountain warfare with the object of rudely disturbing Hitler's forces in Norway. Then they were retrained as air-portable – infantry landing in Dakotas on a captured airfield and deploying very rapidly indeed. Third time lucky, they arrived in September to help the Canadians and the marine commandos to capture the key island of Walcheren, whose occupants were preventing Antwerp harbour from being opened. At 0630hrs on 18 January the 4/5th Battalion Royal Scots Fusiliers of 156th Brigade started from Tuddern, just north-east of Sittard with the objective of Waldfeucht in the wooded, hilly, but also swampy Echterbosch region six miles north-east of Sittard. The Saeffeln Beck, the 20 foot-wide stream with a high bank on one side and marshy patches along its course, was lined with minefields on both sides. No tank, flail, Crocodile, AVRE, Wasp or Weasel could be of help. However, the Canadian Rocket Battery fired four massive salvoes in support, which helped the Fusiliers to cross, despite heavy casualties. The 6th Cameronians had the same problems, but 11,000 smoke shells screened Hongen. Fusilier Dennis Donnini won the Victoria Cross in Stein. Although wounded several times, he threw grenades, rescued another wounded Fusilier, took up a Bren gun and charged the enemy, firing all the time, until he was killed.

Soon Havert, Lindstein and Heilder were captured. The 5th Battalion Highland Infantry suffered forty-three casualties and captured ninety enemy, and the 6th Cameronians suffered equally taking Breberen and Nachbarheid. Bocket fell to 6th HLI; it was the key to open the way to Waldfeucht.

The Lowlanders had the support of the well-trained 8th Armoured Brigade, who were largely frustrated by the streams, minefields and swamps. However, the armour got across a Bailey bridge and 2,000 yards of open, snow-covered country leading to Breberen. Their accurate shooting wrecked the enemy strong points and helped the Lowlanders immensely.

The taking of Waldfeucht was achieved by 155th Brigade under Brigadier MacLaren, but their armoured Kangaroos could not get into major action. 4th KOSB were faced by those rare beasts – Tiger tanks, five of them lurking around Koningsbosch. On 21 January, 5th KOSB encountered them in Echterbosch. Their intrepid anti-tank platoon sited their 6-pdr guns and at close range, 100 yards, destroyed two of them. The Borderers got into Waldfeucht but were heavily counter-attacked. A Tiger tank did a lot of damage to 5th KOSB until 4th KOSB appeared in the evening. There was a bitter little action supporting 5th KOSB of the Lowland Division around Waldfeucht, as a strong German counter-attack came in. Three inches of snow had fallen and three Tiger tanks loomed up in the mist 100 yards away. In the town square the infantry of both sides were slogging it out. A Tiger knocked out a Hussar Sherman and panzerfaust fired from the top floor of houses knocked out two more Shermans. KOSB casualties were heavy. Two more Shermans were knocked out, again by panzerfaust. By midday the enemy had retaken three-quarters of the town and by the end of the day 13/18th Hussars had lost twenty tanks, the last five to concealed 88mm guns.

Within three days, 156th and 157th Brigades had made progress along the central axis, and 155th on the north-west flank.

The task of seizing Heinsberg, a key town of 5,000 inhabitants, was allotted to 155th Brigade. The RAF had flattened it, and 7/9th Royal Scots

Battle for St Joost, 20/21 January, by 1st Rifle Brigade, 8th Hussars and Durham Light Infantry.

Havert, captured in January 1945 by 5th Highland Light Infantry.

The attack on Heinsburg by 4th King's Own Scottish Borderers.

and 4th KOSB advanced by night into the outskirts. When tanks of 8th Armoured and the trusted Crocodiles had arrived by the 24th, they managed to keep and hold the town despite heavy gunfire from the Siegfried Line.

The Lowland Division, which included 'Pontius Pilate's Bodyguard' (the Royal Scots) and 'Hell's Last Issue' (Highland Light Infantry) had done well in appalling battle conditions. The division had lost 101 killed in action, 752 wounded and 258 sick, who were evacuated. The divisional artillery had fired more than 100,000 rounds of 25-pdr ammunition. At Hongen, in Germany, they were the first British division to set up their headquarters in the Third Reich. And for some strange reason a Moscow communiqué of the day credited the Lowlanders with the fifty hamlets, villages and towns captured in Operation Blackcock.

The unhappy Wessex Wyverns, after decimation in Normandy, acute problems in Operation Market Garden, five weeks garrisoning 'the Island', and Operation Clipper, the joint operation to capture Geilenkirchen with the American 84th (Railsplitters) Division, were spending December recovering. In the panic when Wacht am Rhein appeared on 16 December, a small task force went south to help guard the three bridges over the river Meuse.

In early January 1945, Major General Thomas took over temporary command of 30th Corps when Lt General Horrocks went on sick leave to the UK.

Brunssum was the centre of the Wyverns' social activities: NAAFI, ENSA, hot baths (vital!) at the mining pitheads, snowball fights with the Dutch children, even romantic attachments. It was a quiet period before the next storm and certainly the Wessex Wyverns had encountered many storms and were to encounter more. In Operation Blackcock the division was given three objectives: Hart, Jug and Kettle.

Prime Minister Churchill, who had views on every aspect of war and peace, also had them on the codewords for military operations. He forbade any codeword that implied levity! He said that it was humiliating for any of the British Army to risk their lives for a *frivolous* objective. So the Wyverns' objectives had mundane names, but of course they were indisputably dangerous.

130th Brigade was responsible for Operation Hart, the capture of Langbroich and Waldenrath, and for the second part of Jug, 129th Brigade for the first part of Jug, i.e. Wetterach, and 214th Brigade for Kettle, Randerath, on the far eastern flank.

On 20 January, 4th Somersets captured Schummer Quartier and then walked into Langbroich unopposed and 5th Wiltshires took over Breberen from 52nd Lowland. The next day 4th Dorsets travelled in style in armoured Kangaroos against Schier Waldenrath. On the next day the Hampshires took Putt and helped by Crocodiles did a 'pepperpot' barrage and captured Waldenrath. That was the end of Operation Hart.

Operation Jug followed on the 23rd, with 4th Wiltshires starting from Birgden in Kangaroos, dressed in white to blend with the snow – painted gas capes, bedsheets and snow suits – on their way to take Straeten. Their back-up was formidable: Crocodiles, flails, SP A/Tk guns (there were still Tigers roaming around) and Churchill tanks of the Grenadier Guards. 4th Somersets were tasked with taking Scheifendahl, a desolate area of farmland, featureless and covered with a thick carpet of frozen snow. Corporal Doug Proctor noted: 'Each section was to have its own Kangaroo but as usual, chaos reigned. Only sufficient arrived to transport one company [not the whole battalion].' Next came Erpen, over 1,500 yards of open country. Lt Sydney Jary reported: 'An extraordinary sight silhouetted by the setting sun on a ridge about 2,000 yards away was a long column of 150 German troops. Behind us a platoon of

Some of the prisoners captured during operation Blackcock being escorted to the POW cages.

8th Middlesex Vickers MGS firing special Mk VIII long-range ammo engaged them. I could see the Germans scatter in panic.'

Every British division had a specialist regiment, often the Middlesex, the Kensingtons or the Manchesters, who were support troops with 3-inch mortars which could fire a heavier mortar bomb much further than the standard 2-inch mortar. They also had medium machine guns (MMG) which were far superior to the infantry Bren gun. The cheap little unreliable Sten was ideal only for close-quarter fighting.

The 8th Middlesex was the MMG and heavy mortar regiment of the Wessex Wyverns. Corporal Don Linney recalled: 'We set up our Vickers MMGs for a really big barrage. Behind us was a troop of Welsh Guards [in Churchill tanks of 6th Independent Guards tank brigade] who treated us to a rendition of beautiful choral singing with many of the Welsh favourites ending with "Guide me, Oh thou great Jehovah" to the stirring tune Cwm Rhondda.'

The Wyverns on the eastern flank closed in on the river Roer with small actions in Schleiden, Utterath, Aphoven, Schafhausen, Hullhoven, Dremmen and Porsein – all probably mentioned in another Moscow communiqué!

In front of Schafhausen was a wide, flat valley through which the river Roer flowed. It became a No Man's Land with snipers by day and patrols by night.

Lt John McMath, signals officer of 5th Wiltshires, described the classic attack on Utterath:

Carried in Kangaroos under the usual huge barrage; together with the Bren carriers and troop of Churchill tanks there were over 100 armoured vehicles, surged forward over the flat open snow-covered fields. They looked like a mighty fleet of warships. At this fearful spectacle the enemy gave in at once. Utterath was captured, 80 prisoners taken and only one casualty suffered. Then the small villages of Baumen and Berg were taken. [Later] L/Corporal Holloway with his section removed 126 Reigel mines from frozen ground. On 26 January the enemy sent over pornographic leaflets in shells!

Brigadier Essame's 214th Brigade in reserve prodded their way through minefields to Nirm, Kraudorf and Randerath without undue trouble. Hart, Jug and Kettle were accomplished. When Major General Ivo Thomas, the fearsome divisional commander, returned from

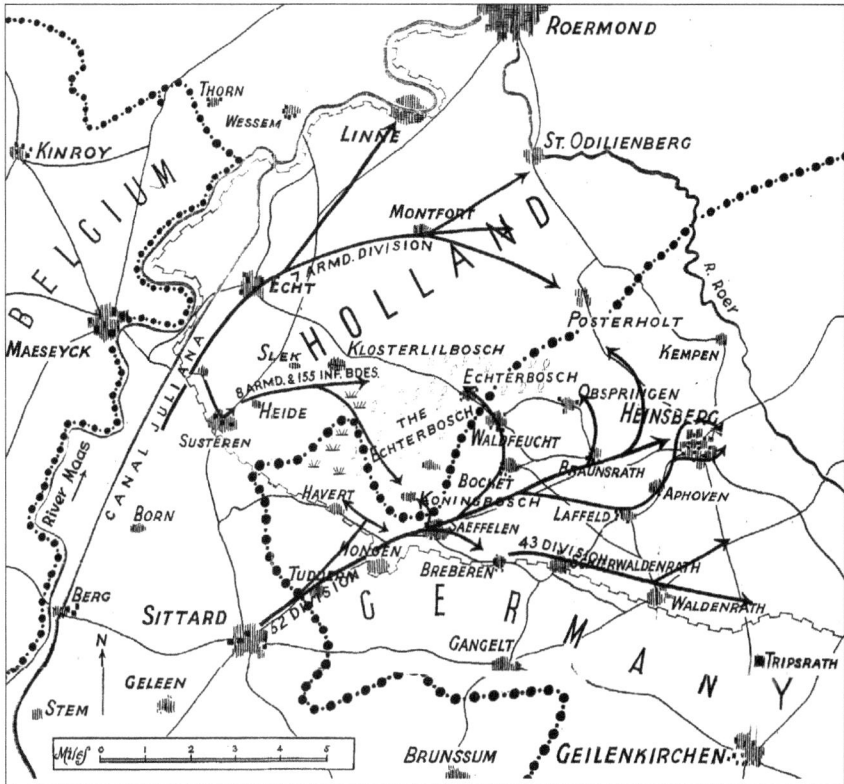

Operation Blackcock.

commanding 30th Corps, he would have been pleased with the way the Wyverns, without him, had made mincemeat of the unfortunate 183rd and 176th Wehrmacht divisions.

The fine armoured regiments of the Red Fox 8th Brigade – 13/18th Hussars, Sherwood Rangers Yeomanry and 4/7th Royal Dragoon Guards – had a very difficult time in Operation Blackcock. They could rarely venture off the icy roads and tracks because they would get bogged down immediately. There were minefields everywhere. Often the bridges would not take the weight of a Sherman. The enemy deployed 88mm anti-tank guns, SP guns, often Tigers, and more frequently, the deadly close-quarter bazooka.

CHAPTER 6

Betuwe: The 'Island' Campaign

In every serious war since time began there has been an area between two opposing armies denied to both sides. In the First World War of 1914–18, No Man's Land was often scores of miles long between one side's barbed wire defences and the other's equivalent defence. For weeks, sometimes months, efforts were made to break the deadlock, always with large, utterly futile casualties. Artillery fire was meant to destroy the wire to allow passage through for the poor bloody infantry (PBI). In the north-west European campaign for over *six months* a No Man's Land existed in the Betuwe. This was the Dutch name for 'the Island', a large rectangle north-south between Nijmegen and Arnhem, two substantial towns ten miles apart. In the centre was the small town of Elst. The Wettering Canal bisected the Island, running some 20 miles west-east through Zetten, Valberg, Homert, just north of Elst and meandering south-east towards Haalderen. The northern boundary was the Lower Rhine running west-east from Wageningen, Randwijk, Renkum, Meteren, Driel, Arnhem, then south past Huissen to the junction with the river Waal, which winds itself through Nijmegen with its two vast bridges, road and rail, west past Oosterhout to Druten, and parallel to the river Maas heading for the North Sea.

The No Man's Land was an area of about 150 square miles and the military issue was simple. The aggressive German forces under Field Marshal Model wished to drive the Allied defenders out of the south of the Island. And come what may they wanted to destroy the vital Nijmegen bridges.

Nearly all the British infantry divisions at one stage or another garrisoned the Island, and for two months post Market Garden so

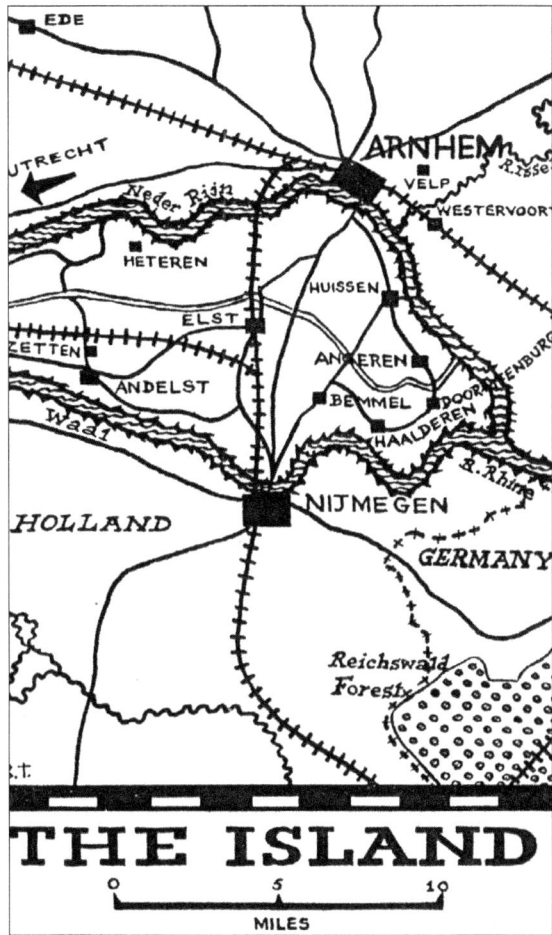

The Island.

did the US 101st Airborne Division. It was no arena for armour as the Guards Division would remember sadly from their Rhine endeavour in September 1944. The ingenious Major General Percy Hobart had naturally designed amphibious war machines. The Buffalo fought and carried infantry, the DUKW carried, and both swam if they had to. They were joined by the Terrapin and Weasel (See Chapter 5).

The Dutch *verdronken* (drowned) land and the manmade polders indicated the very high water table. Slit trenches quickly filled with water. Artillery platforms had to be carefully strengthened and shored up. Farms and civilian houses were at a premium to the defenders, and fairly easy targets for the attackers.

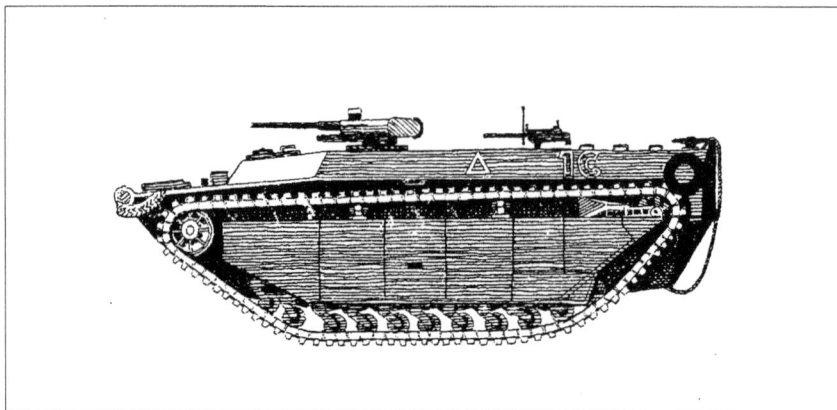

A Buffalo. See page 72.

Captain Athol Stewart described the plight of the PBI in the Dutch polder land:

Do you know what it is like? Of course you don't. You have never slept in a hole in the ground which you had dug while someone tried to kill you … a hole dug as deep as you can as quick as you can … the site given you by an officer who isn't as much interested in your comfort as in putting you where you can kill Germans. It is an open grave and yet graves don't fill up with water. They don't harbour wasps or mosquitoes [as in Normandy] and you don't feel the cold clammy wet that goes into your marrow! At night the infantryman gets some boards, or tin, or an old door and puts it over one end of his slit trench, then he shovels on top of it as much earth as he can scrape up nearby. He sleeps with his head under this, not to keep out the rain but to protect his head and chest from airbursts. Did I say sleeps? Let us say, collapses. You see the infantryman must be awake for one half of the night. The reason is that one half of the troops are on watch and the other half are resting, washing, shaving, writing letters, eating or cleaning weapons; that is if he if not being shot at, shelled, mortared or counter-attacked or if he is not too much exhausted to do it.

When he is mortared or shelled he is deathly afraid and in the day-time he chain smokes, curses, or prays, all of this lying on his belly with his hands under his chest to lessen the pain from the blast. If it is at night smoking is taboo. If there are two in the trench they sit one at each end with their heads between their knees and make inane remarks … such as 'That one landed on 112 Platoon', or they argue as to the type of the shell. If the enemy are coming the soldier has to stay above ground and does not notice the shelling much. He is too busy trying to keep the enemy away. A trench is dug just wide enough for the shoulders, as long as the body and as deep as there is time. It may be occupied for two hours or two weeks. The next time you are near some muddy fields after rain take a look in a ditch. That is where your man lives!

Because of the dominance in the skies of the Allied air force, the Germans became masters of the night. Food, rations, drink and supplies of ammunition came up to the front often in horse-drawn carts. Reinforcements came up and replaced the front-line troops. The wounded were taken back to relative safety and the dead buried. But above all the Germans usually fought better at night than the Allied troops. Their patrols were more frequent and usually more dangerous. They always intended to dominate No Man's Land.

The battle arena of the Island was low-intensity fighting; none of Monty's grand set-piece attacks. A PBI section would encounter a hostile PBI section. The British would radio back for help – quick help – from their divisional MMG and heavy mortar regiment. Captain Michael Bayley was a troop commander with the 2nd Kensingtons and his sketches illustrate perfectly life and death on the Island.

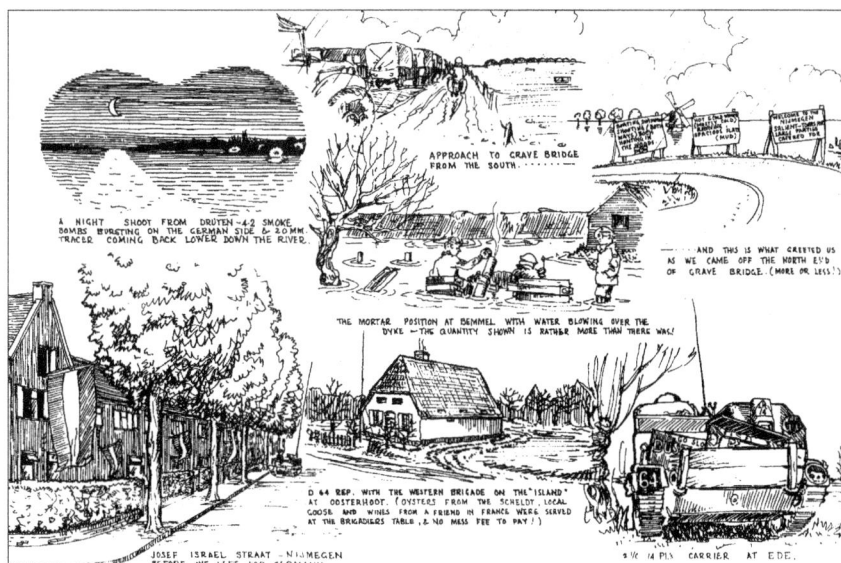

Winter on the Island with the 2nd Kensingtons. (*Michael Bayley*)

Patrol boat in the Dutch polders. (*Imperial War Museum*)

Corporal Barnett's fighting patrol on the Island. (*Imperial War Museum*)

Patrolling activity during the winter months was intense on both sides and there were frequent clashes. There were four types of patrol: *Reconnaissance* patrols were sent out to gain information without fighting for it, including efforts 'to capture a prisoner by stealth' in order to find out details of his unit; *Fighting* patrols were usually of 10–12 men, on occasions up to platoon strength, whose object was to capture or destroy enemy positions and certainly to obtain prisoners and information; *Contact* patrols had to establish communications with 'friendly' flanking units; and *Standing* patrols, up to platoon strength, occupied protective positions forward of the main defences to give warning of impending attacks. Such patrols were prepared to fight, but not at all costs! The Germans were usually highly skilled at having relatively weak forward positions on which the open barrages would fall, i.e. defence in depth.

On 8 December 1944 the 25th German Army in Holland flooded the greater part of the Island, the Betuwe. The 51st Highland Division and the 49th Division, the Polar Bears, were both under the command of General Crerar's Canadian Army. It was always known that from 1 November onwards under conditions of high water in the rivers Waal and Lower Rhine, the enemy could swamp the area by breaching dykes at the eastern end of the Island. So Operation Noah was planned. One contingency was that the Germans would combine flooding with renewed attempts to destroy the great Nijmegen road bridge. Problems would concern the civilian refugees, livestock and military traffic control. On 2 December, dykes were blown north of Elst and also west of Arnhem. Most of the marching troops came off rowing storm-boats, as did civilians, but herds of cattle and civilians had priority and retired southwards in good order.

Field Marshal Model and General Kurt Student's determined army of kampfgruppen, bolstered by experienced young paratroopers, a Dutch SS battalion, and remnants of SS panzer divisions, were determined to recapture Nijmegen if possible and certainly to destroy the huge road bridge there. Much ingenuity was deployed by the British to protect the bridge and by the Germans to destroy it! Floating mines, expert frogmen with explosives, floating boats or rafts with mines or explosives, occasional Luftwaffe sorties and one-man submarines were all deployed. Brockforce was the name of the special force under Lt Colonel D. V. G. Brock, consisting of the 2nd Kensingtons, whose medium machine-gun and 3-inch mortar regiment was a vital part of the Polar Bears Division defending the crucial Nijmegen road bridge. Their rifle marksmen, with great pleasure, shot anything that moved in the river, mostly pieces of wood, wreckage, hen-coops, anything that might harbour an explosive. A comprehensive floating boom was constructed several hundred yards upriver of the bridge, covered by searchlights (known as 'Monty's Moonlight') at night, plus MMG and mortars.

On 13 January 1945, there was much anxiety for Brockforce. The wreckage lying in the river at the railway bridge was shattered by an explosion at 0300 hours. It was an attack by three midget submarines. One beached on the north bank of the river Waal and the crew were killed by shelling. Another was hit by artillery fire and blew up, and the third was heavily engaged, turned round, dived and disappeared. They carried mines or torpedoes, and also large floating logs with mines lashed to them – 12 feet long and 2 feet in diameter. Despite being hit by 40mm shells, the logs did not blow up and the mines had to be defused. Nevertheless, it was a brave effort by the Boche frogmen. The defenders were now christened HMS Brockforce.

Sgt Bill Hudson of the 757th Field Company RE remembered:

> The three RE Field Corps took their turns to operate the 'Woolwich Ferry' across the Waal. It was a Class 9 close support raft running to a scheduled timetable. Lt Oglesby was our 'Skipper'. The main channel was fairly narrow, but about one third of the strip on either side was made over flooded fields and farmland with houses, barns, fence poles protruding out of the water. Our many other jobs included laying booby traps for 2nd Essex, and anti-personnel mines for 2 Glosters, demolishing a footbridge at Wettering canal, dismantling a 50 ft Bailey bridge at Slijk-Ewijk, and constructing a hot baths unit.

Zetten was a medium-sized village six miles north-west of Nijmegen. It was rather unusual, for in mid-January a company of the 1st Leicesters held nine houses at one end and the Germans a dozen houses at the north end. Company HQ had two houses which were linked to each platoon by wireless and line. The gunner FOO and mortar platoon fire controller were both based at company HQ. Suddenly, early in the morning of the 18th, twelve companies of paratroopers – two or three battalions – attacked the Leicesters on their way to capture and destroy the Nijmegen Bridge. They had crossed the Lower Rhine during the night and supported by bazookas, mortars and Spandau fire, a furious battle was started. Reinforcements from the Leicesters, the 2nd Gloucestershires and a troop of Canadian Sherman tanks were sucked into the fighting. Zetten was lost until Major General MacMillan, who had taken over from 'Bubbles' Barker on 30 November, got fed up and on 20 January sent the 2nd South Wales Borderers and the 2nd Essex to recapture Zetten. Altogether the battle for Wetten lasted four days and the Polar Bears had 220 casualties, but they captured 400 paratroopers and accounted for another 300 killed or wounded. The 2nd Kensingtons fired 300,000 mortar bombs in support and the artillery regiments a similar amount. The battle was fought in blinding snow and bitter cold, and cases of frost-bite needed treatment. The Germans fled across the Wettering Canal to the hamlet of Indoornik, which was immediately flattened by Typhoons, artillery and mortars. The 'Nijmegen Home Guard' had done rather well! But the Germans still held the villages of Elden, Driel, Huissen, Heteren and Angeren.

Divisional gunners with 25-pdrs were based on both sides of the river Waal, although flooding of gun pits was a great problem. Their forward observation officers (FOOs) were in the villages of Bemmel, Haalderen, Elst, Valberg, Andelst and Zetten, aloft in church steeples or on the first or second floor of any suitable building. Much of the time the No Man's Land or lagoons were covered in snow or ice and patrols wore snow suits. Patrols of varying strengths laid mines, booby traps and tried to take a prisoner for unit identification. Many patrol clashes were fought at night and 'naval engagements' were fought in flimsy canvas assault boats. The success of a patrol often depended on captured goats, sheep, pigs or chickens!

The usual tour of duty was four weeks in the line and two weeks of rest in Nijmegen. The Dutch families were splendidly hospitable and dances, concerts, cinemas, parties and above all, the bath units, were available.

The winter frost went on until 6 February, and the huge 2nd Army offensive to clear the rivers Maas and Rhine started two days later. Most units disliked the eastern sector around Haalderen. Lt A. A. Vince of the 2nd Essex recalled:

> We agreed that it was the worst sector to look after. On the right ran the river Waal well above ground level and contained by a 'bund'. The only road was under small arms fire and approach had to be made by the Jeep track made out

of rubble from the destroyed houses. Whether the ground was frozen or feet deep in black mud, it was impossible to dig and one had to fight from the ruins of buildings and from cellars. Not a house remained whole and almost every room was fortified with sandbags and chests filled with dirt. In the small slits where windows used to be, there was a weapon of some kind with the safety catch permanently at fire and a finger always on the trigger. The few houses not defended were mined and the unwary patrol from either side rarely left such a building in one piece. Contact with the enemy was as little as 100 yards. The whole battalion area could be, and was, swept by Spandaus and other weapons. The little clusters of wrecked buildings earned their own names such as 'Rotten Row', 'Sniper's Alley' and 'Spandau Joe the Third'.

Lt Colonel Hart Dyke, who commanded the Hallamshires (York and Lancaster Regiment), described his front line between Andelst and Zetten:

> Starting from right to left along the Wettering canal were the isolated farms of Het Slop and de Taart, then Talitha Kumi on the far bank of the river, on the main road Zetten – Indornik – Randwijk. On the left was Hemmen, where there was a castle, two houses and a demolished bridge. The Hallams put out strong patrols on 10 February across the Wettering.

The Polar Bears were ordered to patrol even more offensively to simulate a coming divisional advance, to mislead the enemy, thus drawing off reinforcements from the main 2nd Army battle in the Rhineland and west of the Siegfried Line. Brigadier Gordon ordered the Hallams to occupy the De Hoeven farm, 6,000 yards ahead from Zetten. The Dutch SS troops in front knew their patch only too well and the Hallams had several disasters. One section's assault boat was blown up on a mine causing ten casualties, mostly fatal. Another platoon, twenty-nine strong, of B Company was trapped and all its members killed, wounded or taken prisoner. On 11 March the 2nd SWB relieved the Hallams.

The Island was garrisoned for four months by the Polar Bears who were not required to take part in Operation Veritable and the breaking of the Siegfried Line. Lt Colonel C. D. Hamilton, the CO of the 7th Duke of Wellington's (from Yorkshire) recalled:

> The 'I' staff headed by Captain 'Wrecker' Tris Bax published the first issue of the *Yorkshire Pud* – a digest of world and battalion news, with cartoons which,

through snow or battle, has come out every day since. At the end of February we had another six-day break off the Island at Druten where a taxi service of V-1 bombs overhead made a rest period more harassing than usual. I wonder if GHQ realise that in a 'rest' period as they are so charmingly called, one works twice as hard. We now discovered that 220 days of the 250 since D-Day had been spent in the line. We have taken more Boche prisoners than we had lost in casualties and since D-Day had travelled almost 700 miles. We were still to be the Nijmegen Home Guard.

Fusilier Ken West, a signaller with A Company of the Royal Scots Fusiliers, in his book *An' It's Called a Tam-o'-Shanter*, had a number of anecdotes. Part of his diet included cheese and marmalade sandwiches, which, toasted over a charcoal and cow muck fire, produced a flavour beyond description. He, and others, listened to Glenn Miller on the American Forces network: 'The song "American Patrol" would come over the airwaves and instantly we would hear the machine-gunners firing in time to the music. Brrup, ti, brrup, ti, brrup, brrup and so on right through the song.' Ossie, their cook, a swarthy Geordie of about thirty, had devised a special jerry can cooker. His porridge, bangers, and powdered-egg omelettes and apple tarts were regarded with much favour. They saw a screaming whining plane over Bemmel. 'We could see the sleek silver belly of its fuselage. "It got no bloody propeller an' it's arsehole's on fire," was one inspired comment.' It was not a V-1 or a V-2 but a brand new ME 262, the first operational jet plane in the world. In their pastorie (church) the Scots Fusiliers repaired the smashed organ with tape and chewing-gum. Ben Brenner, a peacetime cinema organist, played 'Bells Across the Meadow', 'Silent Night' and 'The First Noel'. At the Wintergardens, most big American films were shown, including Bing Crosby and Bob Hope in the 'Road' series; Laurence Olivier in *Henry V* was also very popular.

When Captain Murray MC reported a party of fifty-seven enemy passing through an orchard and the brigade stood to, he was afterwards nicknamed 'Heinz'. There were no beans and no enemy, but certainly fifty-seven trees and possibly a goat or two. When the great thaw came, twenty dead Germans and three dead British soldiers were discovered buried deep under the snow.

Every night the Polar Bears listened on the radio to 'Mary of Arnhem', who had a nice cultured voice with a friendly approach. Radio Hilversum gave the war news as seen through German eyes, often mentioning British towns (pubs and dance halls) damaged by V-1 and V-2 bombs.

In April the 'Nijmegen Home Guard' rejoined their comrades of the BLA. In the autumn of 1944, they had liberated the large Dutch towns of Roosendaal, Tilburg and Breda. For the final months of the war the Polar Bears would take a major part in the liberation of northern Holland.

A few miles west of the main Island battleground, the indomitable 11th Hussars, the Cherry Pickers, had spent many dreary, cold, wet weeks in and around Dutch and Belgian farmhouses and hamlets. This is Lt Brett-Smith's description of Moerdijk in early February 1945. It had a world-famous mile-long bridge, road and railway side by side, a great engineering feat, which had been blown up by the retreating Germans at the end of 1944:

Moerdijk was an exception even here, for it was nothing more than a shambles, with not one house untouched by fire, shot or bomb. The only signs of life were the cats – rangy, furtive animals who would suddenly streak across the street from one ruin to another. Of the German occupation there was little enough sign – a destroyed SP on the main street, an AA gun near the convent, and one grave – a rough affair with four empty shell-cases to mark the corners and the usual scrap of paper in a bottle, the cross made by nailing two bits of packing case together. From the tower of the convent you could see as far as the fine church of Dordrecht on a clear day, and all the village spires and windmills and red-roofed hamlets in between stood out in a vivid and colourful landscape when the sun was shining. On the far bank there were a number of white pillboxes manned by the enemy who sometimes got over-confident and sunned themselves on the grass nearby or watched their rations and ammunition coming up in little carts. Occasionally we registered a direct hit on these pillboxes but they were too strongly built to yield to our small mortar-bombs. Still it no doubt shook all the occupants and may have done some harm to the careless ones. But the chief memory of Moerdijk remains the dead village itself – the upturned piano in a front garden, riddled with bullet holes, the grave grey statue of St Francis of Assisi behind the convent gazing unseeingly at a pile of empty bully-tins, an old Ford car lurching on its rims in a charred garage, its tyres nothing but grey ashes. Moerdijk was strangely reminiscent of the villages in Normandy: there was the same smell of death, the same incongruous but terrible destruction.

CHAPTER 7

Reichswald Battles: Operation Veritable

When war was declared in late 1939, a sweet, funny little song became popular: 'We're going to hang out our washing on the Siegfried Line.' Nobody knew for certain where the Siegfried Line was – presumably in Germany facing the formidable French Maginot Line.

The British Liberation Army had always known that sooner or later Monty would ask them to break through the Siegfried Line, which Milton Shulman describes so vividly:

> The ... factor that was instrumental in saving the German armies in the West in the Fall of 1944 was the Siegfried Line. The construction of this well-publicized system of fortifications had begun in 1936 immediately after the reoccupation of the demilitarised Rhineland, and had been feverishly worked upon until the fall of France in 1940. Closely following the 1939 German frontier, the line extended approximately 350 miles from the Swiss border opposite Basle to the junction of the Belgian, Dutch and German borders at München-Gladbach. From there north to Cleve little had been accomplished in the way of deep concrete shelters, and at this end the line rather tapered off into a series of isolated bunkers and field fortifications.
>
> The Siegfried Line varied in depth, strength and effectiveness from place to place along its entire length. Thus, in the area of the Saar where it was at its strongest, it achieved a depth of nearly three miles. Forts were scattered in profusion in this sector, attaining a density of about 40 forts per 1,000 square yards. In contrast to this, the line along the Rhine from Karlsruhe to Basle was only about a half-mile deep and contained only two rows of forts. In the area of Aachen, where the Allies had first reached this belt of fortifications, the Siegfried Line consisted of two thin strings of forts with little density or depth.

The forts themselves were of different designs, but they were usually manned by either machine-gun or anti-tank gun crews, and sited to produce a closely interlocked zone of fire. The roofs and walls were built of cement some five feet thick, and their average size was about 35 feet by 45 feet. The normal complement of men for such forts was about ten, and they lived a damp and cold existence in them. When a battle was at its height, it was considered suicide to vacate one of these bunkers, because Allied artillery and mortar fire was centred on the entrances. Not daring to leave their shelters, even for the purpose of relieving themselves, unbearable sanitary conditions were soon added to the other discomforts of the inmates.

Although no new work had been done on the Siegfried Line after May 1940, and a large amount of the wire had been removed and the mines lifted in the intervening years, this system of fortifications still presented a formidable barrier in the Fall of 1944.

When the Allies crossed the Seine in the early Fall of 1944 engineers and construction troops were rushed to the German frontier in a feverish effort to renovate and improve the long-disused fortifications. From the Fatherland every able-bodied man was hurried forward to be ensconced in the cement bunkers and there they were exhorted to stem the impending invasion of the Reich. The Siegfried Line was now Germany's last hope. On 15 September Field Marshal von Rundstedt issued the following concise order:

1. The Siegfried Line is of decisive importance in the battle for Germany.
2. I order:
   The Siegfried Line and each of its defensive fortresses will be held to the last round and until completely destroyed.
3. This order will be communicated forthwith to all headquarters, military formations, battle commanders and troops.

Note the wording of this order – 'to the last man' had been replaced by the more practical 'to the last round'. And from behind their hastily occupied concrete forts these new troops, fighting well on their own soil, managed to check an immediate invasion of the Fatherland. But it is extremely unlikely that their efforts would have been of much avail had Allied supplies been able to keep up with Allied armoured columns.

The BLA was to be tasked with breaking into and through the northern 14-mile stretch of the Siegfried Line. The first objectives were the two medieval German towns of Cleve and Goch, then through the great Reichswald Forest south to the river Maas, thence to Wesel on the Rhine.

This story does not attempt to do justice to the Canadian divisions of General Crerar's fine army, who fought equally horrible actions in Operation Veritable, many of them nautical, in the battlefields flooded by the German defenders.

Part of General Dwight Eisenhower's 'broad front' strategy – no *schwerpunkt* for him – was for Field Marshal Montgomery's British and Canadian forces to clear and occupy the area between the Rhine and Maas rivers. Planning had started in November 1944. The first plan was called 'Siesta', to clear the Island eastward to the Pannerdensch canal, but it was thwarted when the Germans flooded most of the area. General Horrocks then made another plan, called 'Wyvern', for his 30th Corps, which was handed over to General Crerar's Canadian Army staff. It then became 'Valediction', and was for a south-east advance against two German divisions, the 84th and 190th Wehrmacht infantry. Then Hitler's sudden onslaught in the Ardennes delayed everything and in the New Year 'Valediction' became 'Veritable', which General Horrocks described in *A Full Life*:

> We resumed our interrupted preparations for the Reichswald battle. 30 Corps was lent to the 1st Canadian Army for this operation, designed to destroy all German forces between the Rhine and the Meuse. It was to be a two-pronged affair. We, the northern prong, were to attack in great strength on 8th February. Then, when all the German reserves were on the move north to meet this threat, the southern prong, General Simpson's 9th US Army, would cross the river Roer and advance towards us.
>
> The German forces would thus be caught in a vice. If they elected to fight west of the Rhine they would be destroyed and fewer German troops would be available to counter our thrust into the Reich itself. If they decided to withdraw back over the Rhine, 21st Army Group would be right down on the bank poised to make a crossing anywhere along its length, the primary object of the Rhineland battle.
>
> In theory a perfectly straightforward plan, but not quite so easy as it looked. All operations fought by armies are largely influenced by the shape of the ground and also by weather conditions. None more so than this one.

Above: The 'dragons teeth' of the Westfall – Siegfried Line. (*Harrison Stanley, 111-CC-109298-4-129-46*)

Left: Field Marshal Montgomery, General Horrocks and Prince Bernhard of the Netherlands outside 30th Corps headquarters before the Reichswald Battle, February 1945.

Opposite above: The capture of the Reichswald Forest and Cleve, February 1945.

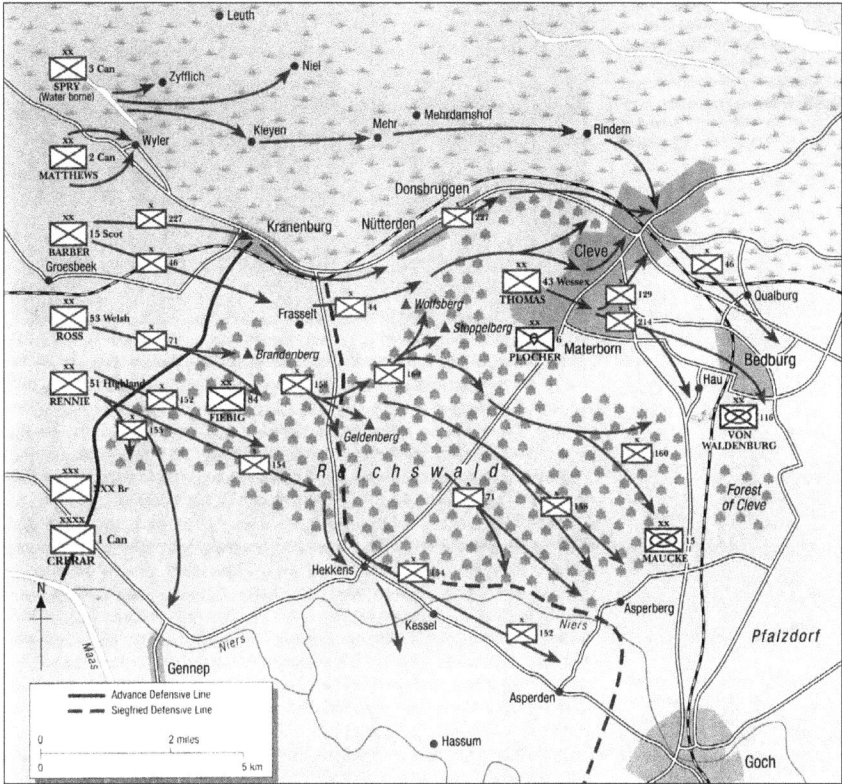

B Company 2nd Argyll & Sutherland Highlanders aboard their 'big friends' in Kranenburg, 8 February 1945, at the start of Operation Veritable, in which 15th Scottish Division suffered over 1,500 casualties.

Troops of 154th Highland Brigade move into the Reichswald on the opening day of the operation. A heavy bombardment of the forest had blasted the positions of the German 84th Division and the going was relatively easy. Resistance stiffened the next day when the enemy bolstered his defences with high-calibre troops. (*Imperial War Museum*)

Churchill tanks bogged down in the mud in the Reichswald, February 1945.

Right: Buffalo with PBI in Kranenburg on their way to Cleve, February 1945.

Below: AVRE supports 53rd Welsh Division night attack in Reichswald Forest, 9 February 1945. (*Birkin Haward*)

43rd Wessex Wyvern Division advancing through Kranenburg to Cleve, 9 February 1945.

Horrocks then went up to the front line and:

I saw in front a gentle valley with small farms rising up on the other side and merging into the sinister blackness of the Reichswald (German forest), intersected by rides but with only one metalled road running through it. North of the forest ran the main road from Nijmegen to Cleve – that is from Holland into Germany. North of this again was the low-lying polder land which had been flooded by the Germans and looked like a large lake with the villages – built on slightly higher ground – standing out above the water. To the north flowed the broad expanse of the Rhine. The Germans were holding the far bank. South of the Reichswald was more low-lying ground which ran down to the river Meuse. This was completely dominated by the southern edge of the forest. The British 2nd Army held the other side of this river. We were therefore faced with a bottle-neck between the forest and the polder land and this had been heavily fortified in depth by the Germans. Moreover the whole area was lousy with mines.

This was really the outpost position of the Siegfried Line which lay 3,500 yards to the east and consisted of an anti-tank ditch, some concrete emplacements, barbed-wire, mines and so on. Furthermore the small Rhineland towns such as Cleve and Goch had been made into hedgehogs, fortresses prepared for all-round defence. The cellars in the houses of all these German frontier towns had been specially constructed for battle – concrete basements with loop-holes and so on, a further example of the careful German preparations for war. Further east still was one more lay-back position called the Hochwald. So the German defences were in considerable depth.

We had to get through this bottle-neck before we could break out into the German plain beyond and the key to the bottle-neck was the high ground at Nutterden. This was the hinge of the door which led to the open country. The front was held by one German division, the 84th, supported by about 100 guns, but we estimated that there were approximately three infantry and two panzer divisions in reserve which could be brought into the battle pretty quickly.

But Crerar and Horrocks were up against all sorts of problems. The first was the great skill of General Alfred Schlemm, who in November 1944 had been transferred from the Italian front to take over from General Kurt Student commanding the 1st Parachute Army. They held a sector of the Western front with four divisions, from the junction of the rivers Rhine and Maas to Roermond. Milton Shulman interviewed all of the surviving German generals at the end of 1945. He wrote:

> Nazi race purists would have had an embarrassing few moments attempting to explain the presence of Alfred Schlemm on an Aryan General Staff. For with his rather short body, his broad Slavic face, his large, bulbous nose and his dark, almost chocolate skin, he looked the antithesis of what Hitler and Rosenberg would have us believe was the true German type. He had been General Student's Chief-of-Staff at Crete, had led a corps at Smolensk and Vitebsk in 1943, and had been responsible for containing the Allied bridgehead at Anzio in Italy in January 1944. His record, coupled with an orderly mind and a keen grasp of tactical problems, placed him amongst the more able generals still available in the Wehrmacht. The contrast between his non-Teutonic physical appearance and his undoubted military ability may not be very significant, but it is interesting.

Schlemm immediately ordered his troops to build a series of defensive lines. The first forward line was held by light troops, whose purpose was to slow down the Allied attack, giving time for his reinforcements to be rushed into place. So in front of the usual thundering artillery barrage that predictably started Operation Veritable on 8 February was General Heinz Fiebig's hapless 84th Infantry Division, which consisted of two regiments: the 1062nd Grenadier and 1051st Grenadier covering the edge of the forest, and the 1052nd Grenadier defending the Rhine flood plain to the north. Two immediate reserves were the Sicherungs

Battalion Munster (elderly Home Guard) and 276th Magen (stomach) Battalion with chronic digestive problems. However, the more powerful 2nd Parachute Regiment defended General Fiebig's sector astride the Hekkens-Nijmegen road, parallel to the little river Niers. General Schlemm's armour was 655th Heavy Anti-tank Battalion with thirty-six assault SP guns. Further back were 180th (Klosterkemper) Wehrmacht Division; Erdmann's 7th Parachute Division, near Geldern; Maucke's 15th Panzer Grenadier Division; the 346th Panzer Loser (Steinmuller Division); and finally General Heinrich Freiherr von Lutwitz's very dangerous 47th Corps. The latter had about ninety tanks between them belonging to 116th Panzer and 15th Panzergrenadier Divisions. Schlemm's second defence line was east of the Reichswald on the axis Rees, Cleve and Goch, and his third and final line ran from Geldern, through the Hochwald via Xanten to Rees. Two miles at the northern end, four miles at the southern end, was the Siegfried Line just east of the only respectable road bisecting the Reichswald.

The second problem was the weather and the effect it had on the battlegrounds, as General Eisenhower reported:

> ... the weather conditions could hardly have been more unfavorable, January had been exceptionally severe, with snow lying on the ground through the month, and when the thaw set in at the beginning of February, the ground became extremely soft and water-logged, while floods spread far and wide in the area over which our advance had been planned to take place. The difficulties thus imposed were immense, and the men had sometimes to fight waist-deep in water ... Under such conditions it was inevitable that our hopes for a rapid break-through should be disappointed, and the fighting soon developed into a bitter slogging match in which the enemy had to be forced back yard by yard.

The BLA had fought on difficult battlefields before: the *bocage* countryside in Normandy, in the Dutch polders on the Island, in the swampy Peel country – but now they were confronted with the huge Reichswald Forest. Colonel Stacey, the Canadian military historian described it:

> Much of the country between the Rhine, Waal and the Maas was open and gently undulating, largely arable, some small woods and suitable for armoured warfare. Just inside the German frontier, the west edge of this rolling plain,

was the Reichswald, a large irregular forested area, eight miles west to east and four miles wide. A dozen miles to the east, the approach to Xanten was barred by the Hochwald and the Balberger Wald forming a belt of woods, one to three miles deep and six miles north to south. The trees in these State forests were young pines four to seven feet apart in rectangular blocks divided by narrow rides. Two paved roads crossed the Reichswald north to south converging on Hekkens. None ran from west to east, so military traffic depended on one-way tracks along sandy rides. The river Niers, swollen by flooding with bridges blown, flowed west along the southern boundary from Goch to Gennes on the Maas.

The logistics for Operation Veritable were formidable. At their peak, the manpower that needed feeding was 476,193, including civilian labour and prisoners. So before the start a stockpile of 2,318,222 rations was built up! For the necessary artillery fire plans 1,471 25-pdr HE shells were needed *per gun* in addition to the 250 rounds carried by each regiment. The RASC brought up huge bulk shipments of petrol. Fifty companies of engineers, three road construction companies and twenty-nine pioneer companies were all fully employed maintaining the few roads and many dirt tracks. These crumbled under the heavy rain, some deliberate hostile flooding and the weight of 30-ton tanks and SP guns.

General Harry Crerar, whose Canadian Army had battled so long and so bravely in the clearance of the Scheldt estuary, which resulted in the opening of the port of Antwerp, now faced a great challenge. Field Marshal Montgomery had entrusted him with temporary command initially with 15th Scottish, 51st Highland and 53rd Welsh Divisions, part of General Horrocks' 30th Corps. The second stage would bring under command 43rd Wessex Wyverns, 52nd Lowland and the Guards Armoured Divisions. The Canadian divisions involved were 3rd Division, who would be waterborne and clear the enemy from isolated fortified villages which had been stranded by rising flood waters on the northern flank, and the 2nd Division, also on the northern flank outside the Reichswald on relatively dry land.

The battleground for Operation Veritable was triangular in shape with Nijmegen as the north-west apex, and Wesel as the principal objective 25 miles east-south-east, via Cleve, Calcar and Xanten. The south-east attack followed the line of the rivers Maas and Niers, with objectives of Goch, Weeze, Kevelaer and Geldern.

Allied information showed that the German reserves which would probably appear during Veritable were 47th Corps under General von Luttwitz. This still-powerful formation included General Plocher's 6th Parachute Division, General von Waldenburg's 116th Panzer Division and General Mauke's 15th Panzergrenadiers. Immediately south was General Meindel's 2nd Parachute Corps with 7th Parachute, 8th Parachute and Fiebig's 84th Wehrmacht. Two more Wehrmacht infantry divisions, the 180th and 190th, were guarding the line south of Weeze to the south of Venlo.

The British Army was outstanding in many ways, for instance its code words for operations and objectives. In Veritable, the planners had decided that Goch was Bangor, Cleve was Wigan, the Reichswald Forest was Repton, and Nuttenden was Ratby. The various report lines were Vindictive, Ramillies, Malaya, Lion, Resolution, Repulse, Temeraire, Valiant, Howe, Kentucky, Virginia etc. Possibly the naval words related to the many flooded battle zones needing amphibious carriers. The official passwords, which everybody forgot, were: D-1 Father Thames, D-Day Virgin Snow, D+1 Silver Ring, D+2 Sour Grapes and D+5 Simple Simon. Just in case Veritable was extended, the spare code words were Plum-Duff and Short-Odds. When 51st Highland Division was advancing on Goch via Hervost, Grafenthal and Asperden, there were six strong pillboxes to be seized, which were called Shem, Ham, Japhet, Faith, Hope and Charity.

General Eisenhower had planned a vast pincer movement. Operation Grenade was an ambitious plan for General Simpson's US 9th Army to capture no fewer than seven dams over the river Roer in the heart of the Westwall/Siegfried Line. But German engineers blew the dam machinery and destroyed the discharge valves. The result was a steady and very powerful flow of water that would create a long-lasting flood along the river Roer, preventing further progress by the Americans. So Field Marshal von Rundstedt immediately sent further divisions north to reinforce General Schlemm's defences and Eisenhower sent two divisions north to replace the 52nd Lowland and 11th Armoured Divisions who joined in Veritable. The Scots became the right (western) flank attack advancing south-east parallel to the river Maas.

It was planned that Operation Veritable should have three separate phases: clearing of the Reichswald and securing the line Gennep-

Asperden-Cleve; capturing the axis Weeze-Udem-Calcar and Emmerich; breaking through the Hochwald 'lay-back' defence lines and securing up to the line Geldern-Xanten.

That still left the bitter triangle from Wesel west to Xanten, south-west to Geldern, with General Alfred Schlemm's troops fiercely defending the approaches into Wesel.

General Horrocks wrote:

> The success of Veritable depended on two things. First, obtaining complete surprise, and secondly on the weather. If the Germans got wind of our attack they would move up their reserves before the battle started. But the weather exerted the biggest influence of all, because the ground was frozen hard, and if only the frost would hold until 9 February, our tanks and motor transport would be able to go everywhere across country without any difficulty. I had no doubt at all that under these conditions we should break out very quickly into the plain beyond and I hoped secretly to bounce one of the bridges over the Rhine.
>
> Surprise would not be easy. An enormous concourse of men, tanks, vehicles and guns had to be moved into the outskirts of Nijmegen and the woods nearby unknown to the Germans. This involved the most intricate staff work. By day the roads must remain empty, showing just an occasional vehicle – the normal traffic, in fact. But as soon as it got dark, feverish activity began. Vehicles, almost nose to tail, came out of their hide-outs and started moving up. Thirty-five thousand vehicles were used to bring up the men and their supplies. One million, three hundred thousand gallons of petrol were required. Five special bridges had to be constructed over the Maas. One hundred miles of road must be made or improved. An intricate traffic control system had to be set up involving 1,600 military police, and each unit was given the most exact timing.

Veritable was eventually fought by most of the BLA. Horrocks' 30th Corps had 200,000 men of war, supported by 1,400 artillery guns – 25-pdrs and mediums plus every type of Hobart's 'Funnies' – Crabs, Crocodiles, AVRE Petards and, for crossing the flooded areas, Kangaroos and Buffaloes. The RAF bombed Cleve and Goch to pieces on the night of 7 February with further heavy raids on Emmerich, Calcar, Udem and Weeze.

General Horrocks continued:

So I decided to attack with five divisions in line, from right to left, 51st Highland, 53rd Welsh, 15th Scottish, 2nd and 3rd Canadian. Behind were the 43rd Wessex and Guards Armoured Divisions ready to pass through and sweep down the Rhineland.

The first essential was to smash through the 84th as quickly as possible and get the high ground, the hinge, before the Germans could bring up their reserves. It was a race for Nutterden, but at the same time I had to clear the Reichswald itself, otherwise the Germans could have concentrated troops there and struck at my communications. Moreover, I wanted the road running through it because I knew how difficult it would be to supply a large modern army with all its complicated needs along one road.

Initially progress of Veritable was slow, partly due to the fact that it rained for the first eight days. But 15th Scottish captured Kranenburg and 53rd Welsh took the Brandenberg 'feature', 300 feet above sea level, on the way to Geldenberg and Wolfsberg. General Fiebig's 84th Division was overwhelmed, 1,200 prisoners were taken, and six out of fourteen battalions were destroyed as 51st Highland advanced deep into the Reichswald Forest.

Lt Colonel B. A. Pearson, CO 8 Royal Scots (14 Brigade) and Lt Colonel de Wilton, 2 Gordons (227 Brigade) consult a map on the way to Cleve, 10 February 1945.

A Churchill tank from 147th Regiment RAC makes a path through the Reichswald on
10 February. The forest was thought to be impenetrable to armour, but tank crews found a
way through the forest by deducing that if the trees were small enough they could be pushed
over and if they were too large for this, then they were wide enough apart to pass between.
The use of armour in the middle of such densely wooded area came as a shock to the
Germans. (*Imperial War Museum*)

Left: the quick-release coupling which retained the fascine in place on the AVRE.
Right: A close up of a Petard mortar with a 'Flying Dustbin' displayed alongside.

A Weasel and DUKW on their way to Cleve through Kranenburg which was flooded by the Germans when they destroyed the Rhine bank further upstream.

Men of the 2nd Gordon Highlanders from 227th Brigade, 15th Scottish Division, wait in their Kangaroos outside the town of Kranenburg for the drive forward to Cleve. (*Imperial War Museum*)

2nd Gordon Highlanders pass knocked-out German AFVs, advancing from Donsbruggen towards Cleve, 11 February 1945.

R. W. Thompson, the *Sunday Times* correspondent, noted: 'The battlefield has become practically a naval action. The tops of houses, the turrets of derelict tanks, smitten tree trunks and branches or telegraph poles and all the fearful garbage of war gave shape to all this desolation of water. Otherwise it might almost have been the sea.' Horrocks' XXX Corps troops referred to themselves as the Water Rats (as opposed to the Desert Rats). At night they looked for islands of mud for their 'laagers'.

The Germans had seeded all the approaches with mines – Tellers for the AFVs and halftracks, and wooden-cased Schu mines, practically undetectable, which would blow a soldier's leg off. Many mines were linked to others and many were booby-trapped with smaller mines buried underneath the Tellers. The SS had a nasty trick of booby trapping dead bodies, theirs and those of the Allies. The author can vouch for this personally as he saw it in the Dutch Peel country.

On the second day, 9 February, as the 51st Highland Division advanced through Breedeweg, north of Grafwegen, another axis advanced through Bruik and a third north of Bruik – all heading into the heart of the Reichswald Forest.

Lt Colonel James Cochran, OC, noted:

One long slog through this forest with desperate and often hand to hand battles by the infantry night after night. The ammunition and supply columns did wonderful work in replenishing the units up one axis, which after the first day's fighting became a sea of mud and water, of signals linesmen working day and night to keep the main axis and its satellite arteries clear ... the great spirits of the Jocks of the infantry battalions which kept them on top of these desperate German soldiers, Hitler Youth and paratroops fighting as infantry and also like demons, never gave them any rest and dislodged them from one defence position after another.

It took the Highland Division three days of hard fighting to clear Gennep, Heyen and Kessel.

Further north, 53rd Welsh pushed two miles east to Stoppelberg, the highest point in the Reichswald, overlooking Materborn. 15th Scottish cleared Schottheide with 6th KOSB in Kangaroos, 2nd Gordon Highlanders were nibbling at Nuttenden, and 8th Royal Scots seized Esperance Hill. Lt Ross Le Mesurier, a brave Canloan officer whose Scout platoon was attached to 5th Camerons wrote:

Without tank support which had become bogged down, machine-gun, snipers and mortars took their toll throughout the morning of the 9th. My head was creased by a bullet, then a piece of shrapnel hit my back. Later on another fragment hit my upper left arm, which stiffened up. The day dragged on, progress was slow and darkness came early. It also started to rain. Our company wireless operator was badly wounded, in great pain, his uncontrolled moans seemed to draw gunfire. My company commander, Major Donald Callender, yelled out, 'For God's sake, Ross, do something'. The wind was blowing towards the Germans, so I cocked a phosphorous grenade to throw it. It was hit by a bullet or a piece of shrapnel and burst in my face which was covered with blobs of burning goo ... I rubbed handfuls of snow mixed with mud on my face to stop the burning. Some Germans began advancing towards us. Our firing forced them to go to ground. A few of us charged them firing from the hip. My Sten gun jammed, so I freed my shovel. They started to run and we chased them. I hit one with the shovel blade in the neck. He hit the ground in a heap. I swung at another but he ducked and it glanced off his shoulder.

A day in the life of an infantry officer ended the next day when the Seaforths relieved the Camerons and Le Mesurier was evacuated to hospital in Ghent. On the Materborn feature the German 7th Parachute Division resisted with great force around Bresserberg. General Horrocks wrote:

> An entirely different type of operation was carried out by the 44th Brigade of the 15th Scottish. Their task was to breach the northern extension of the Siegfried Line, consisting of anti-tank ditches, mine-fields, concrete emplacements and barbed-wire entanglements. Not one single man was on his feet. The officers controlling the artillery fire were in tanks. The leading wave of the assault consisted of tanks with flails in front beating and exploding the mines to clear passages through the mine-fields. Then came tanks carrying bridges and fascines on their backs to form bridges over the anti-tank ditch. The next echelon was flame-throwing tanks to deal with the concrete pill-boxes, and finally infantry in cut-down tanks, i.e. with the top taken off, called Kangaroos.

By the second night Horrocks' 30th Corps had captured 2,700 prisoners and the following day Field Marshal von Rundstedt ordered: 'Cleve must be held at all costs.'

The Taking of Cleve

Eric Codling, 8th Middlesex, noted:

> We had a visit from Major General Thomas [GOC Wyverns] complete with riding breeches, boots and a large peaked cap. He presented Military Medals won and then explained to us that he was in the dog-house with Monty because the last job [the taking of Cleve] we did was a complete balls-up and he hoped *we* would make a better job of things next time. We almost felt sorry for him but at our level we did not know what he was on about. Apparently the Germans withdrew and the two brigades [both from 43rd] ended up firing at each other! Something must have gone horribly wrong for the General to come and commiserate with us!

Lt General Horrocks wrote:

> General Crerar had asked me, "Do you want the town of Cleve taken out?" By taking out he meant, of course, totally destroyed. This is the sort of problem

with which a general in war is constantly faced, and from which there is no escape. Cleve was a lovely, historical Rhineland town. Anne of Cleves, Henry VIII's fifth wife came from there. No doubt a lot of civilians, particularly women and children were living there ... After all the lives of my own troops must come first. So I said "Yes".'

The RAF shattered Cleve and when 129th Brigade leading the Wyverns tried to bypass it by a road to the south, they lost their way in the rubble, fallen buildings and enormous bomb craters, and ended up in the centre of town, in the dark. The fire fight in difficult street fighting at night ended with friends firing vigorously on friends: 129th Brigade Wessex on 214th Wessex and probably Wessex on Scotsmen as well, since 15th Scottish were tasked with capturing Cleve from the west! To be fair to General Horrocks, he admitted his error: 'I unleashed my first reserve, the 43rd Wessex which was to pass through the 15th Scottish and burst out into the plain. This was one of the worst mistakes I made in the war ... caused one of the worst traffic jams of the whole war. In spite of me, the 15th Scottish and the 43rd Wessex forced their way into the shattered ruins that had once been Cleve and some very hard fighting took place.'

Horrocks optimistically had hoped to capture Cleve on the first day in order to wheel southwards and take Goch.

From Monty downwards British generals were nearly always almost grotesquely optimistic with their military plan. It took two days difficult fighting by 15th Scottish to capture Cleve on 11 February. The old, spiteful enemy, 7th Parachute Division under General Erdmann, had turned up to bolster the defences of Cleve and 6th Parachute Division under General Lt Plocher was lurking a few miles away.

Colonel Mutussek, the garrison commander of Goch, the western cornerstone of the Siegfried Line, had 180th and 190th Wehrmacht Infantry Divisions and 2nd Parachute Division under his command – all part of General Straube's 86th Corps. 116th Panzer Division was defending the northern approaches and 15th Panzergrenadier Division was defending the Forest of Cleve, just north-west of Goch. Unfortunately, they put up a staggering defence, which it took four divisions of General Horrocks' 30th Corps to subdue, despite every possible combination of bombardments and death from Hobart's terrible 'Funnies'.

Infantry from the 2nd Seaforth Highlanders of the British 51st Division advance through the Reichswald. In the background a Churchill Crocodile flame-thrower tank rumbles forward ready to deal with a strong point that the battalion encountered during the advance. (*Imperial War Museum*)

The 15th Scottish Division advancing through Cleve, which was severely damaged when Bomber Command attacked the town with Blockbuster bombs instead of the incendiary bombs General Horrocks had requested, February 1945.

The Guards Armoured Division resting prior to the advance on Goch, February 1945.

British infantry preparing to advance to Goch, 16 February 1945.

82nd Assault Squadron RE AVREs with assault bridges and fascines during a night attack, Operation Supercharge, near Goch. (*Birkin Haward*)

A jeep carrying wounded passes through Goch on the way to a First Aid post.

Operation Veritable, 8 February 1945.

Lt General Horrocks, Field Marshal Montgomery and Major-General Barber, GOC 15th Scottish Division, confer during Operation Veritable.

The Oxfordshires and Buckinghamshires move up to do battle in the Reichwald on board a 'big friend'.

German POWs captured in Operation Veritable.

The Capture of Goch

After the capture of Cleve the next few days were spent slowly consolidating and keeping up relentless pressure on General Schlemm's forces, as General Horrocks recalled:

> From now on the battle developed into a slogging match as we inched our way forward through the mud and rain. It became a soldier's battle fought most gallantly by the regimental officers and men under the most ghastly conditions imaginable. It was a slog in which only two things mattered, training and guts, with the key men as always the battalion commanders. The Germans rushed up more guns and more divisions. Eventually we were opposed by more than 1,000 guns, 700 mortars and some ten divisions; they were certainly fighting desperately to prevent our getting to their famous Rhine. Slowly and bitterly we advanced through the mud supported by our superb artillery.
>
> Historians may well in the future pass over this battle as dull but it was far from dull for the front-line soldiers. As I went round day after day I marvelled at the stoicism of these youngsters. They were quite unmoved by the fact that they were the cutting edge of a vast military machine stretching right back through bulging lines of communication to war factories in the United Kingdom.
>
> The strain to which the soldier of to-day is subjected is far, far greater than anything experienced by his grandfather or his great-grandfather.

On 13 February, the sixth day of Operation Veritable, Field Marshal von Rundstedt issued the following order of the day to the German Armies:

> Soldiers of the Western Front.
>
> The enemy is on the march for a general attack on the Rhine and the Ruhr. He is going to try with all the means in his power to break into the Reich in the west and gain control of the Ruhr industry. You know what that signifies so soon after the loss of Upper Silesia. The Wehrmacht would be without weapons, and the home country without coal.
>
> Soldiers, you have beaten the enemy in the great battles of the autumn and the winter. Protect now your German homeland which has worked faithfully for you, for our wives and our children in face of the threat of foreign tyranny. Keep off the menace to the rear of the struggling Eastern front, so that it can break the Bolshevist onslaught and liberate again the German territory in the East.

My valiant fellow-combatants. The coming battles are going to be very hard but they demand that we stake our utmost. Through your perseverance the general attack of the enemy must be shattered. With unshakeable confidence we gather round the Führer to guard our people and our state from a destiny of horror.

VON RUNDSTEDT

Field Marshal

General Major Wolfgang Mauke's 15th Panzergrenadier Division was positioned south of Cleve between the square Forest of Cleve and the Reichswald. General Major Siegfried von Waldenburg's 116th Panzer Division moved towards Bedburg east-south-east of Cleve and was soon locked in battle with Thomas' 43rd Wessex Wyverns, who had taken Trippenberg. 4th Wiltshires lost 200 casualties in three days fighting. Eric Codling, 8th Middlesex, was there. 'Next task to secure escarpment overlooking Goch from the north. A bitter struggle against a fanatical enemy. DF rained down and continued for what seemed an eternity. Soon the German fire shifted to the slowly advancing infantry, slowly advancing but much reduced in numbers. Difficult to climb out of our refuge – a sure sign of "bomb-happiness". First real evidence of long-term effects of months of action whittling away the reserves of courage that we had enjoyed at the beginning of the campaign. The bank balance was running out.' On the 15th, 4th Dorsets confronted Kampfgruppe Hutze. Wally Caines was the radio operator in the scout car of the adjutant, Captain Richards:

> 1000hrs objective three small villages on a crest overlooking the outskirts of Goch. The German paras opened up with all they had. Spandaus and every sort of weapon. They had dug into the ground floor of the houses in the first village, concealed themselves excellently. Shells and mortars rained down and both company COs, Majors Symonds and Gill, were wounded in the afternoon. By now Jerry was falling back. Very few prisoners were taken. The boys had no intentions over worrying about that.

Colonel George Taylor, the much-decorated CO of 5th Duke of Cornwall's Light Infantry, noted: 'four battalion attack. 1st Worcesters on the right, 7 Somersets on the left, 5 DCLI and 4 Somersets would exploit southwards to get on the high ground overlooking Goch.'

Starting from the hamlets of Blacknik and Berkhovel, General Thomas planned to capture Pfalzdorf, Imigshof, Bermanshof and Schroenshof and then to cut the main road from Goch north-east to Calcar. General Horrocks wrote:

> The turning-point came on 16 February, when the 43rd Wessex Division carried out a brilliant 8,000 yards advance which brought them to the escarpment overlooking the fortified town of Goch, which was subsequently captured by the Jocks of the 51st Highland and 15th Scottish Divisions. At this time my right flank was very much a Scottish army. The 52nd (Lowland Division) were on the extreme right and one of the regiments supporting the attack on Goch was that magnificent medium artillery regiment The Scottish Horse.
>
> On the right of the 15th Scottish were the 53rd Welsh Division who disappeared into the Reichswald itself where they were to spend one of the most unpleasant weeks of the whole war owing to the difficulty of movement up the narrow boggy rides. [One of them was called 'Chewing Gum Alley', which Churchill would have vetoed!] They fought their way steadily forward against increasing German opposition. This was the division which, in my opinion, suffered most. For nine days without ceasing they edged their way ahead. The narrow rides made tank support almost impossible and they were constantly being faced by fresh German reserves. [The battle for the Asperberg feature by the Niers river two miles north of Goch lasted two whole days, 14–16 February.]
>
> They never faltered and on 18 February they reached their objective, the east edge of the forest, having suffered 5,000 casualties – 50 percent of the total lost by this division during all these operations in Europe. The 18th of February is a day of which Wales has every right to be proud.

Alastair Borthwick describes in his book *Sans Peur, 5th Seaforth in WW II* how the 51st Highland Division attacked Goch through Asperden: 'We watched as several mattresses of rockets went over, each weighing ten pounds, equivalent to a medium 5.5-inch shell. They made a sound like rushing water as three mattresses each with 300 rockets passed overhead into Asperden.' The enormously thick 60-foot square concrete pillboxes in the Siegfried Line were heaped over with earth. 'Infantry and supporting arms dealt with the trenches outside. The casement was the weak point but even 17-pdr anti-tank shells simply bounced off. But AVRE Churchill tanks with a 40-lb "dustbin" bombard thrown by a petard would blow the steel shutters out of the casemate. Then a Crocodile would come up, squirt unignited fuel through the casemate for half a minute, and then fire one ignited squirt after the rest. The garrison died instantly and horribly. Quite soon the German defenders realised that their supposedly "impregnable" Siegfried Line pillboxes had become a death trap.'

The Allied air forces and artillery concentrations had destroyed most of Goch. The streets were blocked to tanks and wheeled traffic and the house debris was ideal cover for snipers, Spandaus and bazooka teams. The enemy were well dug-in, in gardens, behind buildings and in cellars. 1st Gordons suffered 140 casualties, 5/7th Gordons 133, Argyles 121, Black Watch 125, 2nd Seaforth 141.

On 21 February Veritable was over – all the objectives, including Goch, captured.

During Veritable the Canadians earned two Victoria Crosses (Sgt Aubrey Cousins and Major Tiltson). They lost 379 officers and 4,935 men as well as many non-battle casualties. But General Horrocks' 30th Corps lost 770 officers and 9,600 men. Most were lost in the final week of February. The general hospitals in Louvain, Brussels and Bruges were crammed full with the casualties.

Lt General Horrocks wrote:

> During the course of this horrible battle nine British and Canadian divisions supported by a vast array of artillery had been under command of 30 Corps. We had smashed our way through carefully prepared enemy defensive positions under the most unpleasant conditions possible. No one in his senses would wish to fight a winter campaign in the flood plains of north-western Europe, but there was no alternative. We had encountered and defeated three panzer, four parachute and four German infantry divisions. We took 16,800 German prisoners. Enemy casualties were about 75,000 against 15,634 suffered by us. This was unquestionably the grimmest battle in which I took part in WWII.

The third volume of Colonel C. P. Stacey's official history of the Canadian Army, *The Victory Campaign*, notes: 'Let no one misconceive the severity of the fighting during those final months. In this, the twilight of the gods, the defenders of the Reich displayed the recklessness of fanaticism and the courage of despair. In the contests west of the Rhine in particular they fought with special ferocity, rendering the battle of the Reichswald and the Hochwald grimly memorable in the annals of this war.'

The poet Alan Seeger wrote in his poem '*Rendezvous*': 'But I've a rendezvous with Death / at midnight in some flaming town'. In and around Goch, Welshmen, Scots and Wessex Wyverns died in the terrible killing grounds of Veritable – the largest, longest and most awful battle that the British Army of Liberation fought inside the Third Reich.

Field Marshal Montgomery, General Horrocks and Whistler after Operation Veritable. (*Imperial War Museum*)

Captured German kampfgruppen troops. (*Imperial War Museum*)

CHAPTER 8

Reaching the Rhine:
Operation Blockbuster

General Alfred Schlemm commanded the 1st Parachute Army and by 20 February had repeated the tactics he had used so successfully in Italy at the Anzio beachhead. There he had built up a force from nothing to eight divisions in a few weeks. His force to defend the important Wesel pocket guarding the river Rhine was the remnants of Fiebig's 84th Infantry, Klosterkemper's 180th Infantry, and Hammer's 190th Infantry. He had three armoured formations in place: the Panzer Lehr, reformed after their mauling in Normandy; Maucke's 15th Panzergrenadiers; and von Waldenburg's 116th Panzer Division. But his best troops were the 6th, 7th (Erdmann) and 8th (Wadehn) Parachute Divisions. They had taken a beating in the defence of Goch and the surrounding villages. Milton Shulman described them:

> The resistance of the parachute divisions was as grim and relentless as anything yet seen in the West. In these young, indoctrinated Nazis, fresh from a Luftwaffe that had ceased to exist, the faith in their Führer and in their own cause had not yet died. They had been rushed into the role of infantry troops following the Allied victory in Normandy. Not having personally felt the sickening impact of defeat, they had not yet given way to the despair and hopelessness that by now had gripped most of the Germans fighting in the Siegfried Line.
>
> The parachutists in the Reichswald were that in name only. They had never been taught to jump from an aeroplane. The bulk of them had received no more than three months' infantry training. But they possessed two other compensating virtues – youth and faith. Over 75 percent of them were under 25 years of age, and together with the SS the parachutists contained the best remaining products of young German manhood. And they were nurtured on

Left: German paratroop machine-gunner.

Below: Detail of picture on page 36. (*Imperial War Museum*)

the deeds of their predecessors – the men who had landed at Crete, broken the Maginot Line at Sedan, and held Cassino. They developed an *esprit de corps* which the regular army divisions had long since lost. When they took their places in these new parachute divisions, hastily being formed to take advantage of this spirit, they were given speeches such as this, delivered by Lt Colonel von der Heydte to the men who had just been sent to his regiment: I demand of every soldier the renunciation of all personal wishes. Whoever swears on the Prussian flag has no right to personal possessions! From the moment he enlists in the paratroops and comes to my regiment, every soldier enters the new order of humanity and gives up everything he possessed before and which is outside the new order. There is only one law henceforth for him – the law of our unit. He must abjure every weaker facet of his own character, all personal ambition, every personal desire. From the renunciation of the individual, the true personality of the soldier arises. Every member of the regiment must know what he is fighting for.

He must be quite convinced that this struggle is a struggle for the existence of the whole German nation and that no other ending of this battle is possible than that of the victory of German arms … He must learn to believe in victory even when at certain moments logical thinking scarcely makes a German victory seem possible … Only the soldier who is schooled in philosophy and believes in his political faith implicitly can fight as this war demands that he shall fight. This is the secret of the success of the Waffen SS and of the Red Army – and lack of this faith is the reason why so many German infantry divisions have been destroyed.

Colonel von der Heydte was an outstanding German soldier who had fought in the Cotentin peninsula in Normandy, commanded a kampfgruppe with some success against 30th Corps in Market Garden and was determined to keep on fighting until the bitter end. His boss, General Schlemm, not only faced grave dangers from in front as he defended the Wesel pocket, but also from behind! His orders were simple – to hold. 'Once the battle was joined,' said the general, 'it was obvious that I no longer had a free hand in the conduct of the defence. My orders were that under no circumstances was any land between the Maas and the Rhine to be given up without the permission of the Commander-in-Chief West, von Rundstedt, who in turn first had to ask Hitler. For every withdrawal that I was forced to make due to an Allied attack, I had to send back a detailed explanation.'

Hitler Jugend leave their mark!

Hitler was on von Rundstedt's back, desperate to keep the shipping traffic going along the Rhine for passage of coal from the Ruhr mines, which was shipped along the river to the Lippe Canal, south of Wesel, and from there up the Dortmund-Ems Canal to the northern ports of Hamburg, Bremen and Wilhelmshaven. It was the lifeline for the German Navy, to keep its large U-boat fleet supplied.

Hitler also ordered Schlemm not in any circumstances to blow up any of the nine bridges over the Rhine in his command sector without the Führer's permission. Schlemm moved his HQ from Emmerich to Rheinberg, where he had direct radio communication with each bridge commander. Moreover Hitler threatened Schlemm that if a bridge were captured intact he would answer with his head. No excuses or explanations of any kind would be accepted.

The German 1st Parachute Army had available 717 minnenwerfer mortars, 1,054 guns including 50 dual purpose 88mm guns and a considerable number of SP assault guns. Field Marshal Montgomery wrote, 'the volume of fire from enemy weapons was the heaviest which had been met so far by British troops in the campaign'.

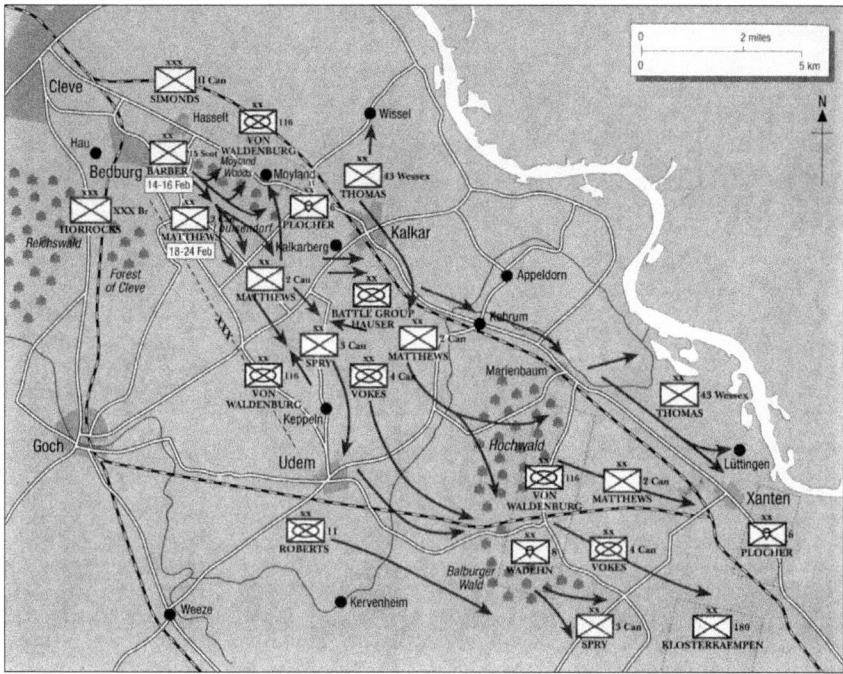

Operation Blockbuster, February 1945.

Operation Grenade had been delayed by the four-foot rise in the water level of the Rhine, caused by deliberate flooding. The original attack by General Simpson's 9th US Army was scheduled for 10 February but was delayed until the 23rd. The German 15th Army, with four weak infantry divisions, was soon in trouble and Schlemm was forced to send two armoured divisions – Panzer Lehr and 15th Panzer – plus 406th Wehrmacht Infantry Division south to help. Nevertheless, by the 23rd, Schlemm had stabilised his front, holding a 20-mile line from the Maas to the Rhine, which forced General Horrocks and the Canadians to engage in Blockbuster 1, what General Eisenhower described as 'some of the fiercest fighting of the war'.

Three British formations now joined Horrocks' huge 30th Corps, Guards Armoured, 11th Armoured and 3rd British. For three days 53rd Welsh had some rest out of the line around Nijmegen, 43rd Wessex Wyverns rested in and around shattered Cleve, 51st Highland around Goch, and 15th Scottish went back to the Tilburg area.

To get to Wesel, the key town of Xanten seven miles to the west had to be captured. The three towns of Weeze, Kevelaer and Geldern, all due west of Xanten, were heavily fortified and the densely wooded Hochwald running laterally north-south for three miles was a formidable barrier. The escarpment that stretched from Calcar to Udem was strongly defended by Kampfgruppe Hauser, 116th Panzer Division and 8th Parachute Division.

53rd Welsh, emerging after their week in the icy jungle of the Reichswald, led the advance in Blockbuster, from Goch following the river Niers towards Geldern via Weeze. During their few days of rest in Holland, over 2,000 reinforcements had to be integrated – an immense task. 4th RWF needed 365 men; 7th RWF 359; 6th RWF 276; 4th Welch 281; 1st HLI 239; 2nd Mons 243; East Lancs 148; 1/5th Welch 137 and the Ox and Bucks 203. In a way these young men from HAA, LAA, anti-tank regiments and conscripts were fortunate. What lay ahead of them, apart from surviving Blockbuster, were eight or nine weeks of 'interesting' fighting, a whole series of small town, small village actions with a casualty rate perhaps a tenth of the Normandy bloodbath. But the PBI, the 'grunts', are always at risk in the very front line. Integration was always very difficult for the rookies and for the old sweats.

Operation Leek was the plan to capture Weeze, then Hees and Wemb five miles south-east of Goch, defended by 15th Panzergrenadier and 7th Parachute Divisions with six battered but still ferocious enemy battalions. 71st Brigade would lead, backed by 8th Armoured Brigade (Red Fox's Mask), a squadron of Crocodile flame-throwers and another of flail tanks. Under a massive bombardment of twelve field regiments and seventy medium and heavy guns, 2nd Mons and 6th RWF led and for two days it was a bloody stalemate with heavy casualties on both sides. Corporal Bert Isherwood of 13th Field Dressing Station noted in his diary: 'Sunday 25th. The battle rages forward. 53 Div is having a hard time, large numbers of casualties are coming back, the FA and the FDSs are busy. The way forward and the way back is littered with dead and the debris of war. A shambles of destruction, the near distance thick with smoke and shot through with the flashes of the guns and the blasts of incessant explosions. An infantry supply NCO tells me that the Div has about fought itself to a standstill before Weeze.' It was St David's Day with leeks being bravely worn on every helmet. One of many enemy counter-attacks was led by three Royal Tiger tanks, which were almost impregnable.

During the night of 1–2 March, the enemy survivors pulled out in the dark, leaving a ghost town of mines, boobytraps, utter destruction and desolation.

Above: The 53rd Welsh Division on objective on the east side of the Reichswald, February 1945.

Right: A mortar team from 2nd Battalion Middlesex Regiment fires its 4.2 inch (107mm) mortar in support of British 3rd Division during the attack on Kervenheim. The 4.2 inch (107mm) mortar was one of the heavy infantry weapons provided by a machine-gun battalion to each infantry division. The Middlesex Regiment supplied heavy machine-guns and mortars via the four of its battalions who were attached to the 3rd, 15th, 43rd and 51st Infantry Divisions throughout the war in north-west Europe. (*Imperial War Museum*)

A few miles east, Major General 'Bolo' Whistler's 3rd British 'Ironside' Division was a reinforcement division for Horrocks. Operation Heather was the plan parallel to 53rd Welsh to capture Kervenheim and Winnekendonk, south of Udem, south-east of Weeze. Having relieved 15th Scottish, 3rd British concentrated north of Goch. Captain Marcus Cunliffe, adjutant of 2nd Warwicks wrote: 'Still more enemy units were brought in to fight the northern battle. The effect was comparable to that of the Normandy campaign where British and Canadian formations attracted to themselves the main weight of enemy opposition, thus simplifying the subsequent American advance.' Operation Heather was aptly named. The two divisional centre lines were soft, saturated sandy tracks through woods, heaths of heather and no metalled roads. A nightmare for the heavy Churchill tanks of 6th Guards Tank Brigade and Hobart's flails and Crocodiles. Their first objective was securing the Udem-Weeze road and the vital Muhlen-Fleuth Bridge. When that was taken the 53rd Welsh could attack Weeze from a different flank on the 28th. In the woods, the South Lancs and East Yorks of 8th Brigade had a torrid time in Operation Donald. The German 7th Parachute Division and part of 190th Wehrmacht fought like devils in and around the hamlets of Geurtzhof, Kampshof and Bussenhof. The Corps artillery was firing stonks and barrages against counter-attacks by 1st Battalion 24th Parachute Regiment and 3rd Battalion 22nd Parachute Regiment with support from a few Tiger tanks. Lt Colonel Bill Renison, CO East Yorks wrote: 'It had been what I think Wellington would have described as a "close run thing" but we had won. At a cost of 33 KIA, 123 wounded and 4 missing. We took 150 German prisoners and counted 85 dead of Para Group Hubner in the battle area.' Major General 'Bolo' Whistler, GOC 3rd British, ordered the sappers to place a name board on the bridge with the East Yorkshire's crest and name – 'Yorkshire Bridge'. More or less the same thing happened a mile east where 1st KOSB, 2nd Royal Ulster Rifles and 2nd Lincolns were taking on 1st Paratroop Regiment in Operation Daffodil. They cut the Udem-Weeze road, but tried and failed to get a bridgehead over the Muhlen-Fleuth. The third brigade, 185th, was tasked with the capture of Kervenheim and then Winnekdonk, fortified villages on the minor road south from Udem. 2nd KSLI, 2nd Warwicks and 1st Royal Norfolks took it in turns through woods, waterlogged clay fields, scattered houses under shelling, through minefields and often drenching rain. The young Pte Joseph Stokes of 2nd

Men from the 1st Battalion Herefordshire Regiment cross the anti-tank ditch that surrounds Udem. The ditches were a common type of temporary defence which were thrown up throughout the whole of the Siegfried Line. They caused little difficulty to the infantry and could be bridged in seconds with specialised armoured bridging equipment. Covering fire, smoke and a determined attack meant that troops and tanks could be over the ditch in a very short space of time. The secret was in trying to keep the enemy occupied while the breach was made. (*Imperial War Museum*)

KSLI won the Victoria Cross in the attack on Kervenheim. He personally captured twenty-one prisoners, was wounded eight times, refused to go back to the RAP and eventually fell leading a charge 20 yards from the enemy.

Later, Terence Cuneo painted a fine scene of Stokes' dramatic action, which is now owned by the trustees of the King's Shropshire Light Infantry. The Royal Norfolks suffered 161 casualties capturing Muserhof, Murmannshof and around Kervenheim. Jack Harrod in his book *History 2nd Battalion Lincolnshires* described the final attack on Winnekendonk:

The Bn surged forward under pitiless fire. Many fell including Major Clarke wounded by a grenade. The road junction near the village was reached. Still the enemy parachutists fought back grimly. Snipers fired from first floor windows and Spandaus shot through loopholes in the walls at ground level. Now the light was going fast and the infantry and the 3rd Scots Guards tanks went into the village in billows of smoke, punctuated by the orange flashes of the enemy 88s and criss-crossed by lines of tracer. It was a great and terrible spectacle. The Bn had really got its teeth in and was not to be denied.

Savage fighting continued until the parachutists had had enough. Thirty were killed, fifty were wounded and eighty surrendered. When the village was finally cleared in the morning seventy more gave themselves up. The Lincolns had 107 casualties. Major General Whistler's diary recorded:

We had had our most successful battle. Damn difficult country – centre line non-existent or on a mud track – yet we had done everything asked. We have captured 1,200 prisoners, nearly all paratroopers. We must have killed and wounded many more. The scenes of devastation in Germany are quite remarkable. There is nothing that has not been damaged or destroyed. We have captured Kervenheim, Winnekendonk and Kapellen of the larger places. I would say the 9 Brigade under Rennie have done best, 185 next under Mark Matthews. 8 Brigade have not had so much luck though East Yorks had a terrific night holding a bridgehead against repeated counter attacks … Am getting on very well with Jorrocks [Lt General Horrocks]. He is a fine leader and well liked, quite ruthless but one has to be … The Boche are beaten pretty badly here. We are through the Siegfried Line and very near the Rhine. I would not have missed this battle for anything – unpleasant and frightening though it has been. To fight on German soil after all this time is more than I ever expected. Now for the Rhine.

Further north it took the Canadians five days to break through the Hochwald, Balburgerwald and also capture Udem against General Schlemm's Battlegroup Hauser, von Waldenburg's 116th Panzer and Waden's 8th Parachute Regiment. On their left and northern flank Major General Thomas and the Wessex Wyverns, using the main Cleve-Xanten road as their axis, relentlessly advanced south-east of Calcar, capturing Kehrum, Marienbaum, Vynen and Wardt and reached Luttingen, almost in the suburbs of Xanten.

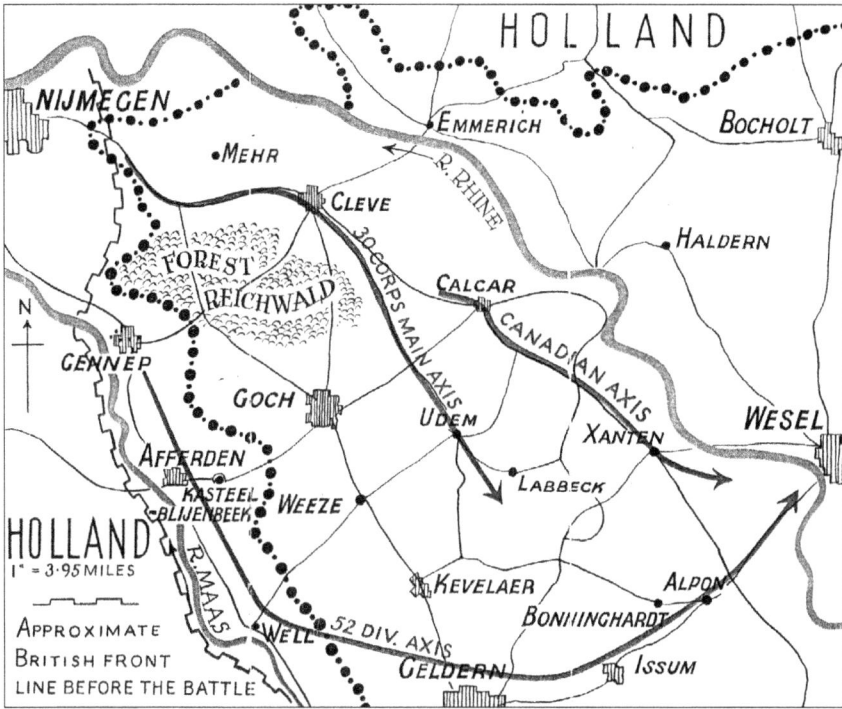

The advance to Wesel.

Blockbuster 2 was the name of the operation to capture Xanten and Wesel. In his excellent book, *The Victory Campaign*, Colonel C. P. Stacey wrote: 'The series of extraordinary orders from Berlin ... as a result the diminishing bridgehead became cluttered with damaged tanks, transport, artillery without ammunition and all the debris of an army fighting a heavy losing action.'

In the Canadian northern sector between Marienbaum and Udem, in fierce fighting, Major F. A. Tilston, the Essex Scottish, won the VC on 1 March, despite wounds which cost him both legs.

General Schlemm's troops were fighting tooth and nail to hold the key lateral road west and in front of Wesel. The line was Xanten, Veen and Alpen. Colonel Stacey pointed out: 'Xanten in history a Roman town, was in German legend the home of Siegfried with a population usually of 5,000, but now crammed full of retreating Huns.'

Because of the limited space, Blockbuster 2 relied on the 2nd Canadian Infantry Division and the 43rd Wessex Wyverns to remove the cork from the bottle. On the afternoon of 3 March, the British 4/7th Royal Dragoon Guards ahead of the Welsh, met the American 17th Cavalry Recce cars in Berendonk, three miles north-east of Geldern. And 52nd Lowland on the western flank passed Wemb south of Weeze and concentrated between Geldern and the river Maas.

On 4 March, Prime Minister Winston Churchill and Field Marshals Alan Brooke, the CIGS, and Montgomery visited General Crerar, the Canadian GOC, and together they drove to the Reichswald and the Siegfried Line. They lunched with General Simonds and visited British 30th Corps afterwards.

To the south of the Balbergerwald, 11th Armoured joined in the fight to push through the Schlieffen Line's Hochwald Layback defences, which ran from Xanten south-west to Sonsbeck and the Winkelsher Busch.

General Thomas had the doubtful honour of mopping up the west bank of the Rhine. 4th Somerset Light Infantry were ordered to capture Xanten and 5th Wiltshires to capture Luttingen. Generals Crerar and Horrocks had several days to get the 'chessmen' in position – hundreds of guns and Hobart's ferocious 'funnies', including Crocodiles, Crabs and AVREs. The 2nd Canadian Division would seize the western sector of

Night attack on Weeze: 82nd Assault Squadron bridge A/TK ditch, supporting the 53rd Welsh Division, 28 February 1945. (*Birkin Haward*)

Monty's moonlight RE assault bridges and fascines support the 51st Highland Division near Goch, 28 February 1945. (*Birkin Haward*)

Skid Bailey bridge launched in night attack on Weeze by 82nd Assault Squadron RE, 28 February – 1 March 1945. (*Birkin Haward*)

Xanten and the high ground to the south. All the attacks for Blockbuster 2 started at 0500hrs on 8 March. Milton Shulman described the scene:

> Schlemm was forced to contract his bridgehead under the mounting pressure of the British, Canadian (and to the south the American) forces surrounding him. By 8 March the bridgehead at Wesel quivered rather uncomfortably in an area about 15 square miles in size. In this postage-stamp sector, because of Hitler's mad refusal to permit the withdrawal of any fit men, there were jammed no less than nine divisions, three corps headquarters and an army headquarters. So crowded were these staffs that one sugar refinery alone housed the headquarters of three divisions. Here these troops who had fought an exhausting battle for a full month without a rest, were pounded relentlessly from the air and from the massed guns of the tireless Allied artillery.

On 6 March, the German High Command grudgingly gave permission for their bridgehead to be evacuated by the 10th and General Schlemm controlled the final operation at 2nd Parachute Corps HQ. General Meindl commanded the rearguard of the remnants of 6th, 7th and 8th Parachute Divisions, 116th Panzer, and 346th Infantry Division. Their only conduits to relative safety were the road and rail bridges at Wesel and numerous ferries.

Schlemm ensured that Xanten was well defended and by the time that the Wyverns and Canadians had captured the town, Wesel itself was practically denuded. The end came on 10 March when both bridges over the Rhine at Wesel crashed into the river, leaving just a small rearguard behind. Many of the houses were inscribed *SIEG ODER SIBERIEN* (VICTORY OR SIBERIA) or *LIEBER TODT WIE TYRANNEI* (BETTER DEATH THAN TYRANNY). The gallant German survivors crossed the Rhine to fight another day.

Operations Blockbuster 1 and 2 finished on 10 March. In Veritable and Blockbuster 1 and 2, the 1st Canadian Army, with its eight British divisions, had a total casualty list of 1,049 officers and 14,585 other ranks. The Canadians suffered 8,942 casualties and the British 6,694. General Schlemm at one point had command of 84th, 180th and 190th Infantry Divisions, 15th Panzergrenadier Division, 116th Panzer, Panzer Lehr Division and 2nd, 6th, 7th and 8th Parachute Divisions – a substantial army of ten divisions. They lost 22,239 prisoners and had estimated casualties of 22,000. The US 9th Army suffered 7,300 casualties in seventeen days of Operation Grenade.

23rd Hussars passing through Wesel in their brand new Comet tanks.

Men of the 1st Battalion Cheshire Regiment landing from Buffalo craft near Wesel, 24 March.

Field Marshal Sir Alan Brooke, Monty, Churchill and General Simpson on the Siegfried Line.

The 21st Army Group, after a month of continuous fighting – and the Parachute Army was the best on Hitler's western Front – were lined up along the western banks of the river Rhine. Hitler sacked von Rundstedt and appointed Field Marshal Albert Kesselring, who had achieved defensive miracles in Italy, to defend the Third Reich.

General Horrocks has the last word:

> This was the grimmest battle in which I took part during the war. No one in their senses would choose to fight a winter campaign in the flooded plains and dense pinewoods of Northern Europe, but there was no alternative. We had to clear the western bank of the Rhine if we were to enter Germany in strength and finish off the war. Although our losses seemed very high, I kept on reminding myself that there had been 50,000 casualties during the first morning of the Battle of the Somme in the First World War. Eisenhower summed up the situation when he wrote in a letter to Crerar, 'Probably no assault in this war has been conducted in more appalling conditions than was this one'.

CHAPTER 9

Capturing Arnhem: Operations Destroyer and Anger

The British Army had suffered a defeat in September 1944, in the Witches Cauldron of Arnhem and Oosterbeek, and revenge was always on the agenda. However, to achieve this, the capture of the desolate, watery No Man's Land of the Betuwe, the so-called 'Island' between Nijmegen and Arnhem, was required. For over six months the British Army took it in turns to garrison the half-dozen or so wartorn villages and hamlets in the south sector while the Germans did the same with the northern sector. Frequently the Germans sent ground, air, water or underwater attacks to try to destroy the Nijmegen road and rail bridges. The 49th West Riding (Polar Bear) Division, commanded in early 1945 by Major General S. B. Rawlins, was nicknamed 'The Nijmegen Home Guard', as they spent more time on the Island than any other formation. The Canadian generals Crerar and Simmonds in the long, bitter clearance of the river Scheldt, which led to the reopening of Antwerp harbour, had jointly recommended to SHAEF that the capture of the high ground between Arnhem and Apeldoorn, 20 miles north, would be most helpful in the planned operations Veritable, Blockbuster, Plunder and Varsity. Christened Operation Anger, the plan was given to Lt General Charles Foulkes of the 1st Canadian Corps, which had just arrived from the Italian campaign. But hostile flooding possibilities of the Island area in March and April were deemed too dangerous. Since the classic capture of Le Havre in September, the Polar Bears had frequently come under the Canadian command.

There were three phases: Operation Destroyer was the overall plan to advance to Apeldoorn. The initial clearance of the south-east sector of 'the Island' was Operation Anger and then 5th Canadian Division

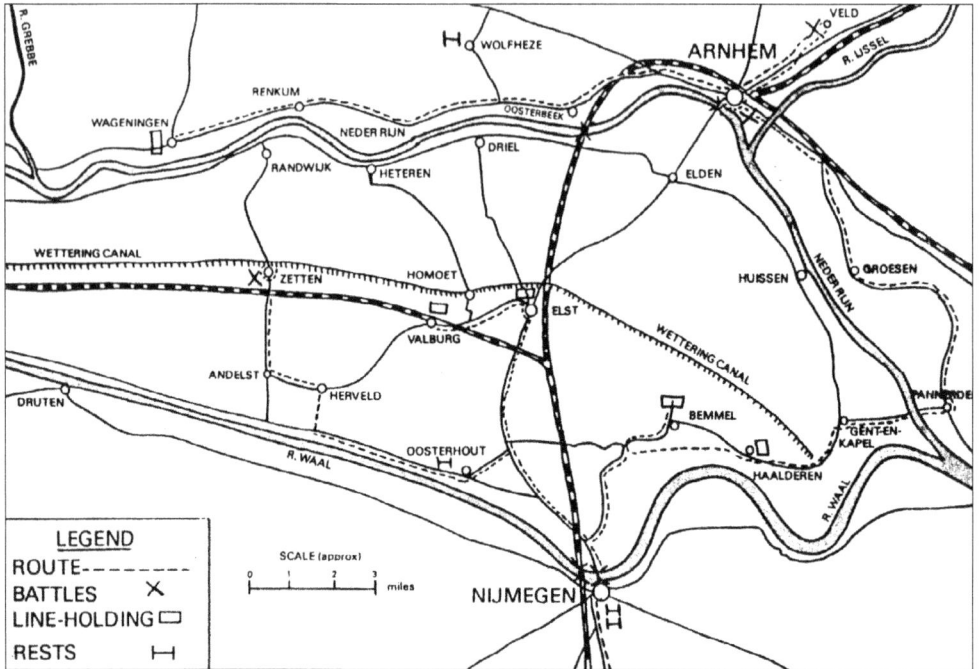

Betuwe – the 'Island' between Nijmegen and Arnhem.

Regional emblem of Arnhem. (*Market Garden Veterans Association*)

would clear the remainder of 'the Island' and cross the Lower Rhine at Oosterbeek by a 'scramble' attack, and then push on to capture Arnhem itself in Operation Quick Anger.

The residents of Arnhem and Oosterbeek (over 450 of whom had been killed in the September battle) had been evicted from their homes, which were then systematically looted of anything of value to aid refugees in Germany. The shattered towns were then turned into strong defensive positions to resist future Allied advances. Arnhem had been extensively shelled over the winter and the Arnhem road bridge that the British had fought so hard for had been bombed by the Allies in October 1944 to deny its use to the Germans. In retaliation for a Dutch railway workers strike supposed to aid the Allies' September advance, the Germans banned all inland freight movement. This prevented food being grown in the north from reaching the south and west of the country and caused thousands of deaths among the Dutch population, in what became known as the *Hongerwinter*.

German forces in the Netherlands had been redesignated as Festung Holland and approximately 10,000 troops of the German 30th Corps under General Philipp Kleffel were holding the region Arnhem-Apeldoorn. The 361st Volksgrenadier Division was west of Arnhem and the 1,000-strong garrison of the heavily fortified Arnhem/Oosterbeek objective included elements of 346th Wehrmacht Infantry Division, 858th Grenadier Regiment, paratroops and Dutch SS troops, the Landstorm Nederland, 83rd (Dutch) 55th Regiment and 34th SS Volunteer Grenadier Division. On the day of the start of Operation Anger, Heinrich Himmler decreed that all cities occupied by German troops should be defended at any price, failure to do so being punishable by death.

Operation Destroyer started at 0600hrs on 2 April. Brigadier H. Wood, CO 147th Infantry Brigade, sent the 1/7th Duke of Wellington's Regiment to clear Haalderen and Doornenburg under a huge barrage from five field regiments, four medium, two heavy, two HAA and a mattress of 240 rockets plus a deadly assortment of 'Funnies' and a call on a RAF cab rank. Bemmel church tower was used as an OP and Haalderen, Gent and Doornenburg were bombarded with 125 rounds of H/E per gun. The Royal Scots Fusiliers and the 1st Leicesters secured the right flank, reaching the Lower Rhine later in the day. Brigadier D. S. Gordon's 146th Brigade and 1/4th KOYLI effected a bridgehead over the

A 2nd Kensington carrier on the Island. (*Michael Bayley*)

1st Leicesters on the Rhine en route to Arnhem. (*Charles Pell*)

Wettering canal and cleared Zand. A mile or so to the east, 4th Lincolns seized Angeren and pushed a further three miles to Huissen, where they were rudely attacked by the RAF, who caused casualties. By nightfall, Huissen was captured and the river Rhine reached. At first light on 3 April, the Hallamshire Battalion (York and Lancaster) took Rijkers Waard and fought their way up the autobahn as far as Kranenburg. The 1st Lothian and Border Yeomanry flails led the Polar Bears from Zand to Rikerswaard, but three 'crabs' blew up on mines.

The Hallams occupied Elden and sent a patrol across the Lower Rhine and 4th Lincolns crossed the river Rhine near Loovier. Another patrol penetrated to the river Ijssel and had a good view of Arnhem from Westervoort on the railway line. By 5 April, Operation Destroyer's first phase was completed. At long last the polder battleground of the Island was in British hands. The new GOC, Major General Rawlins, was very pleased.

Operation Quick Anger

Although patrols had probed the outer defences of Arnhem across the river on 3 and 4 April, the next major operation planned with the 1st Canadian Division was to take place on the eastern flank. The west was overlooked by commanding heights and the crossing would have been difficult and expensive in lives. The tower of the Saxen Weimar barracks had a particularly good view. On 8 April, it was decided that the attack would take place east of Arnhem near Westervoort. Twenty-four hours after the landing of the Canadians near Velp, C Squadron of the Recce Regiment nipped across the river Rhine at Emmerich to look at the west bank of the Lower Rhine and the river Ijssel. The logistical problems were immense. Troops were on the move everywhere, and tanks, trucks, carriers and jeeps congested the centre lines. The Polar Bears moved over a period of several days to the Westervoort area, via the Lower Rhine or via Emmerich. The KOYLI moved on the 10th to Zevenaar. The 56th Brigade were to make the night crossing on the 12th with the Glosters leading to form a bridgehead over the river Ijssel with the 2nd South Wales Borderers and the 2nd Essex following up and advancing into Arnhem.

During the day, Typhoons flew violent raids on all of the German positions. The artillery barrage started at 2040hrs with a two-hour

2nd Battalion, the Essex Regiment, Arnhem, April 1945. (*Harry Conn*)

Pepperpot programme. The 11th RTR's Buffaloes would ferry the attack across the river Ijssel from Westervoort. The Royal Navy's Force U would provide several LCMs (Landing Craft Mechanised) and Canadian engineers had built a pre-fab Bailey bridge upstream at Doornenburg, which would be floated down river to Westervoort just before the attack. The sappers would also build and operate a ferry and the RASC run DUKWs across the river. It was a huge task to get over a thousand troops over a wide river at night under fire.

At 2040hrs on a cold, clear night, aided by searchlights, all four Gloster rifle companies set off. Unfortunately Sod's Law was working overtime. The sapper charges that were set to explode the 'bund' riverbank in front of A and C Companies did not detonate. B Company had trouble with their assault craft engines. D Company went ahead into violent opposition but took the 'Spit', which was their objective. In the centre, A Company and C Company manhandled three of their craft over the bund, crowded into them and sailed across to capture an old Dutch fort and a silk factory that the Germans had fortified. B Company got over a bit late, but took their brickworks objective. Captain R. V. Cartwright of A Company captured sixty fully armed Germans cowering in the inner cells of Scheisprong Fort, almost single-handedly. The Glosters suffered thirty-two casualties, but by 0300hrs the 2nd Glosters, the

South Wales Borderers and the 2nd Essex had passed through according to plan, having been ferried across the river all through the night by Buffaloes and DUKWs – a total of 260 loads to get 56th Brigade across. Lt A. A. Vince of 2nd Essex recalled:

> We raced westwards through the heart of the town, but on the high ground surrounding Arnhem, the Boche had masses of artillery and mortars, so had little difficulty dropping his shells in the correct places. Captain Peter Butler died of his wounds, one of a number of casualties. We took 145 prisoners, the remainder fled westwards leaving incendiaries to fire the town. We saw the evidence of the tragedy of Sept '44, the broken guns and equipment, the little shallow slits the Airborne had dug in a few seconds and from which they had fought for days. We saw the little white crosses in corners of Dutch gardens, often with an inscription such as '31 Unknown British Soldiers'. On top of the cross would be a weather-stained Red Beret placed there by the Germans as a tribute to the cream of fighting men.

Pte H. Fensome of the 13th Platoon 2nd Essex, a nineteen-year-old reinforcement, went into his first major attack well briefed by Sgt Conn, his platoon sergeant, a veteran of D-Day. He remembered:

> We crossed in Buffalo assault boats, landed in the factory area and came under heavy shelling from 88s, rockets and mortar fire. Some of our section got caught in the open. When the barrage stopped Sgt Conn and Corporal Frank Hudspeth went on a recce and found Hoppy Hopton, Olly Oliver and Paddy McCulloch – all unharmed. We then moved into Arnhem clearing the area around the railway station and up a street near the museum. We took eight prisoners in one building. Amongst them was a blonde woman who we thought was 'Mary of Arnhem', the radio broadcaster. They were taken away to the POW cage. During one advance Lt Willis banged on the turret of a Sherman tank with his Sten gun, asking for covering fire. I observed Billy Bebber towing a toy brass cannon taken from the museum, for a souvenir.

By nightfall the 2nd South Wales Borderers were firmly into the centre of the town, most of it badly shattered, and they claimed to be the first battalion into Arnhem. The following morning the sappers bridged the river with prefab Bailey pontoons and the 146th Brigade passed through into Arnhem. Lt Colonel Hart Dyke wrote: 'Friday 13th was a most satisfactory day for the Hallams, with three objectives captured. Next morning the advance continued with A Company and tanks overcoming all opposition. In it Lt Davies was killed and 12 ORs wounded. This brought our total losses up to 158 killed and 689 all ranks wounded or a little more than the total strength of the Bn when it arrived in Normandy 11 months before.' Major Godfrey Harland of the KOYLI recalled:

> At first light with the aid of some Canadian tanks we attacked some German positions in the grounds of Sonsbeek hospital. It was to be our last company attack in NW Europe. The company was digging in, having taken the objective, when the customary German counter-attack by shell and mortar fire arrived, landing on a section of the right hand platoon. Three were killed and two were wounded. Sad to lose those three splendid young soldiers, Geordie Alcock, M. Durham and E. Lees who had fought so many battles with the company only to lose their lives so late in the war.

Detail of picture on page 72. (*Imperial War Museum*)

Detail of picture
on page 152.
(*Harry Conn*)

The 4th Lincolns had an all-day battle to clear the huge ENKA BV factory occupied by the 346th Engineers Battalion and the 46th Festungs Machine-gun Battalion. Tanks, armoured bulldozers, AVREs and flame-throwers were needed. The place was so vast that more Germans kept appearing from all sides to replace those already killed. Eventually, 234 prisoners were rounded up and as many were killed in the fierce fighting. The Fife and Forfars' Crocodiles, the AVRE bombards of 617th Assault Squadron and the Lothians' Crabs crossed after 56th Brigade and were invaluable in helping clear Arnhem in the next two days. The Lincolns suffered fifty-three casualties. The 147th Brigade moved from the Island to join the rest of the division on the night of 13 April. The 7th Dukes passed through the South Wales Borderers towards the railway station. Lt Colonel Hamilton, their colonel, wrote:

Hugh Le Mesurier appeared leading my 'rear' B Company and A Company's route had been blocked by the detonation of the enemy MG Bn. Soon my party were lost in a maze of streets. The company commanders treated every new horror as a great joke. By dawn the bottleneck of roads, Arnhem station and the first high ground was ours. The next day – the 14th – the Dukes joined in a brigade movement to secure all the high ground. The Boche reaction was confusing. He plastered our area unmercifully for half an hour, killing brave CSM Fellows of A Company. At dusk a company of [German] infantry

155

supported by three French Renault tanks tried to re-occupy the crossroads. Prisoners said they had no idea we were there. D Company made short work of the complete outfit. The following day – the 15th – Gerald Fancourt's C Company occupied Arnhem Zoo. Fancourt was soon on the air offering the brigadier a live polar bear to replace the wooden one on his caravan!

The Leicesters had come down the river all the way from Nijmegen to Arnhem using the 36th LCAs of the 552nd Flotilla, got to the top of the Westervoortedijk near the harbour and dug in near the Elisabeth Hospital in the western suburbs. The civilian population of Arnhem had been evacuated in late September with the exception of policemen, firemen and the Landstorm Nederland of the 83rd Dutch 55th Regiment. The latter were dug in around the Schelmseweg-Kempenbergerweg crossroads and gave the Dukes some trouble. The Canadians and General Rawlins were keeping up maximum pressure. To the east, Bronbeck and then Velp were soon taken. On the night of 15/16 April, helped by Fife and Forfar flails, the 56th Brigade led again. The 2nd Essex attacked on the north side of Velp, taking 100 prisoners. The 2nd South Wales Borderers forced their way into the residential area. The Glosters captured the Meteor factory on the Ijssel and took fifty prisoners. The Lincolns passed through towards De Steeg, and Rheden was reached by midday on the 16th. The leading tanks and companies of the South Wales Borderers were amazed to be greeted in Velp by a delirious and overwhelming crowd of civilians, with swarms of excited children, all waving flags, throwing flowers and shouting 'Good boy, Tommy'. It was truly an amazing and exhilarating experience, so great was the relief of the Dutch to be freed from their oppressors.

The left flank of 146th Brigade was protected by the Recce Regiment. Major Gooch wrote in their history: 'We had a field day comparable with our best in France (operating from Terlet) taking 143 prisoners, killing many others and clearing the whole flank of the division.' The Hallams and the KOYLI had advanced on Rozendaal, captured it with the Ontario Regiment Shermans, and thundered on via Beekbergen, Zijpenberg and Postbank to De Steeg. The Hallams' B Company, on tanks, moved on to Dieren and Laag Soeren, where they linked up with other Canadian troops. By the end of 16 April, Operation Quick Anger was over, and Arnhem had been taken at a cost of sixty-two killed and 134 wounded. More than 1,600 Germans had been made prisoner and double that number had been put out of action.

1/4th KOYLI officers at De Steeg, north of Arnhem. Left to right: Major Harland (OC A Company), Major Singleton (C Company), Major Barlow Poole (Bn 2i/c), Major Whitworth (D Company), Major Carter (FOO 69 Field Regt RA), Lt Firth (Intelligence Officer). (*Lt Colonel G. Barker Harland MC*)

Matthew Halton, Canadian Broadcasting Company (CBC), reported:

16 April 1945: Arnhem. Today I drove into the smoking shell of that town, once one of the most beautiful in Holland. There was nothing there now but ruin, and a memory.

It was fitting that it should be a British formation, the 49th West Riding Division, that should take Arnhem at long last and write the words 'Paid in full' across another page of British history. Last September the world stopped breathing to watch this town. If the British Army had been able to link hands with the British 1st Airborne Division, which had landed round Arnhem, the Rhine would have been turned while the German armies were disorganised, and the armoured divisions would have poured into the plains of Hanover and Westphalia. But the effort was a bit too much for us, the weather was against us, and in fact we didn't have enough strength at the decisive place.

I drove in from the south today, past powerful German forts and redoubts which had been shelled and bombed and burnt out by thousands of our rockets firing hundreds at a time. The town was a deserted, burning shell. I visited a British regiment, and saw about 200 German prisoners. Fires were blazing. Machine-guns chattered from the high ground north of Arnhem, and two or three German shells whistled into the town. An airplane engine, that had fallen from some British bomber, disintegrating high above, lay in a little park beside a canal. The whole thing was a dreary, disheartening sight – another of the destroyed towns of a beautiful continent.

There were craters everywhere. White tapes through the minefields. A British soldier who'd lost a foot on a Schu mine. Two other soldiers being buried. Engineers making demolitions. Bulldozers, backing up a few feet, shaking their heads in roaring anger, and then tearing into the side of a broken house. A Martian ant hill – a bedlam of men at war. Motor cycles bouncing back and forth. Tanks and Bren-carriers dusting through to the next battle. Convoy leaders going crazy to get their convoys through. All the noisy, clanking machines and paraphernalia of war. A lone Dutchman, the first civilian we had encountered, came slowly down a long street. He shook hands. 'You have come back', he said quietly. Just that. The British had come back, as they always do. Operation Quick Anger was over.

Operation Dutch Cleanser

The 147th Brigade now received orders to launch an attack on Ede, ten miles north-west of Arnhem. The 146th Brigade, together with the South Wales Borderers and the 2nd Essex, had concentrated around Arnhem and the KOYLI remained in Rozendaal. Rex Flower recalled: 'Billeted with local schoolmaster and family. Four days clearing up jobs. Found a sanatorium, many of the patients had been killed or wounded. Found a mass grave, miserable and brooding. We did not dally. We moved to Wolfeze woods, scores and scores of Airborne gliders, some wrecked, most undamaged.'

Ede was held by 300 Dutch SS troops of the 83rd SS Grenadier Regiment, part of the 34th SS Division. They were a motley unit of renegades, but their fighting spirit and fanaticism had already been proven. On the afternoon of 16 April, the Royal Scots Fusiliers led, together with tanks of the Canadian Calgary Regiment. Progress was slow due to defended German roadblocks with anti-tank guns. The following morning, under a thick smokescreen, the advance continued until halted by the Simon Stevin barracks defences. A heavy artillery bombardment of the barracks at midday also unfortunately caused damage to Ede town centre. Now Wasp flame-throwers of the Royal

Scots Fusiliers set the barracks alight. B and C Companies advanced on the railway station where they quenched their thirst, among an elated crowd in the Stationsweg. Fusilier Ken West remembered: 'We were in the beautiful Veluwe district north of Arnhem. Thick woods of silver birch, oak and beech, a haven for wild life.' He had watched his leading platoon run shouting and screaming up the road and out of sight. On top of a very small hill the fusiliers were dancing around and singing 'I'm the king of the castle' until Major Weir, Webley pistol in hand, appeared. 'It's not a damned Sunday School outing – bloody fools,' he said. Pte Bob Day of the 1st Leicesters recalled:

> In woods outside the town of Arnhem which the retreating Germans shelled heavily as we marched in single file through the street, our company was mistaken by other British troops for the enemy. There followed a rocket barrage, the ferocity of which defies description. When the shrieking and blasts of those devilish weapons came to a merciful end, a young officer's arm was left hanging from the shoulder by a shred. I gave the Bren gunner a gentle nudge (I was his No. 2). He fell to one side limply. He was dead. A piece of shrapnel had penetrated his spine killing him instantly. When darkness came we buried him in the dell with a makeshift wooden cross.

On the 17th, the 2nd Essex returned to Arnhem, embarked on LCAs and LCMs and sailed down the river to Renkum. Lt A. A. Vince recalled:

> We disembarked in this completely razed village and marched into Wageningen again without opposition. Here Lt Colonel Butler DSO left us to return to command a Bn of his own regiment. Lt Colonel E. S. Scott MBE took command. Between our own positions and those of the enemy ran the floodable valley of the river Grebbe and once more the battle became static with extensive patrolling by both sides in which the enemy suffered far more heavily than we did.

Fusilier Ken West watched rocket-firing 'Tiffies' screaming down on the Ede barracks, and C Squadron of the Recce Regiment pushed ahead in front of the 11th Royal Scots Fusiliers to Lunteren. West noted: 'Newspaper reporters and Canadian news-reel men were out in force to record the scenes for posterity. It was rather nice to be the centre of attention. Teenage girls ran to place garlands of laurel around our necks as we marched along singing the old Boer's trekking song which had become a favourite of ours.' The Burgomaster assumed the fusiliers were Canadians. West continued: 'We had to inform him that we were indeed British. "If you were English you would be singing 'Tipperary'." So we sang 'Tipperary' again and again and again.'

The Leicesters had advanced to Ede-South, following the railway line much of the way. On the large Ginkel Heath near Ede the 7th Dukes found much airborne equipment and parachutes from the September landings. Lt Colonel Hamilton recalled:

> There was no respite. At 5 o'clock we followed the RSF out of Arnhem along the great autobahn leading west to Rotterdam. By mid-day Ede was captured and, mounted on tanks, C Company pushed south six miles to the Rhine, thus taking 56th Brigade's objective! Their brigadier gave me a cool reception at Wageningen that night! The Boche had gone back five miles to the Grebbe line and our startling advance was halted while negotiations opened for the passage of food convoys through the Boche line for the starving people of NW Holland.

Operation Dutch Cleanser was a full-scale offensive by the Canadian Army to force a wedge from the Arnhem area to the Zeider Zee to cut off 120,000 German forces. On the evening of the 17 April a halt was called. It was now decided that the 1st Canadian Corps would advance no further into north-west Holland in an effort to prevent the enemy flooding the whole countryside west of Utrecht. However, aspects of war continued. Rex Flower of the 1/4th KOYLI wrote on 21 April:

> Assault pioneer platoon lifting German Teller mines, hundreds of them – but their own Pioneer officer Lt J. C. Crabtree, driver Private J. W. Dean were both killed when their vehicle struck a mine. The last casualties of the war. A truce arranged. Seyss-Inquart, the Nazi CO, threatened if the advance continued he would open the sea dykes and flood the countryside. This threat was taken seriously by our Army Commander. No movement whatsoever during the day on pain of court-martial. But A Company killed, wounded and captured part of a German patrol as they tried to get into the company position. The Germans broke the truce, we didn't. They had no chance.

On the 22 April, the South Wales Borderers left to join the 53rd Welsh Division for the last fortnight of fighting in north-west Germany. It was a

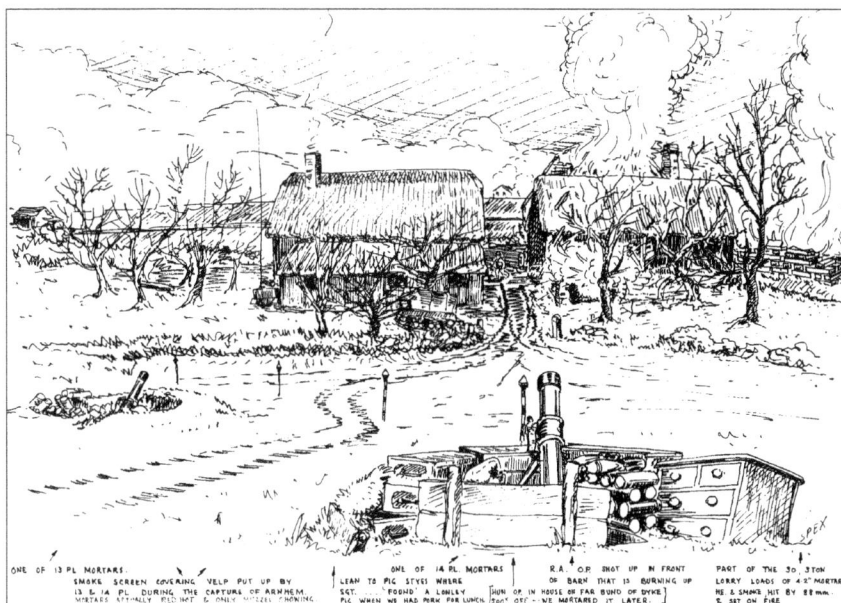

A 2nd Middlesex mortar position near Velp. (*Michael Bayley*)

pleasant surprise for the 4th Lincolns at Wolfheze when their regimental band unexpectedly arrived from Lincoln and played at concerts and church parades. On the 26th the battalion marched to Lunteren, played in by the band as Brigadier Gordon took the salute. Three days earlier the 1st Leicesters had their last bitter little battle at Renswoude against German-Dutch SS troops. Eversley Belfield, General Horrocks' co-author, wrote in his diary:

> ... the vast smoke-screens, a feature of all the river crossings made observation often very difficult. The unhappy town of Arnhem was burning hard with smoke towering up to several thousand feet ... To the right was the smaller town of Velp where I saw people running as smoke shells burst above it starting fires. I wondered whether they were civilians ... The process of liberation is often a most unpleasant one: if we were conquering the Germans I should not care so much, but the unfortunate Holland had suffered enough already. [Later by jeep] I passed into the area where the Airborne Division made their final stand. It is a terrifying sight with wrecked houses, damaged trees, torn and slivered by the mortaring and shelling ... piles of equipment lie around, burnt and abandoned ... The whole area devoid of civilians breathes an air of tragedy.

CHAPTER 10

Weapons of Revenge

Adolf Hitler, that dangerous little corporal, was obsessed with secret weapons of mass destruction. In 1939 his friend and architect Albert Speer took charge of the Peenemünde site for the development of rockets. From 1937 to 1940, the German Wehrmacht spent 550 million German marks on the development of a large rocket, hampered by Hitler's 'divide and rule' policy, which involved not only the three branches of the armed forces but also the SS; even the postal system came in on research facilities! General von Brauchitsch had presented plans to Hitler in 1939 for a *long-range* rocket but this had been rejected.

On Christmas Eve 1944 an experimental V-1 was successfully launched from Peenemünde. The Polish underground movement learned enough for the War Cabinet in London to be mildly alarmed. In May 1943, an RAF reconnaissance plane located the Peenemünde research station and air photographs revealed that pilotless jet-propelled aircraft were being developed.

Two months later a V-1 test made a flight of 243 kilometres without deviating more than one kilometre from its course. On 10 July, Hitler ordered Speer to devote special attention to the production of *vergeltunswaffen* (weapons of revenge). Ninety-six sites along the Channel coast were to be ready by 15 December, with production of 5,000 V-1s a month, and a stock of 5,000 to start the offensive; but in August the RAF inflicted serious damage at Peenemunde and the first batches were faulty and useless.

By the end of November, sixty-three sites had been discovered by Allied flight reconnaissance, and on 5 December Allied bombers began attacking the sites. By February 1944, seventy-three out of the ninety-six

A V-2 for Antwerp seen rising from north Holland over the Nijmegen Bridge.

sites had been severely damaged and the Germans decided to abandon the whole project.

A special unit, Flak Regiment 155, was set up to carry out future launching operations and to comb out all non-German elements from the site construction gangs. Because of the bombing of the production sites, sufficient stocks would not be available until mid-1944.

In parallel, from 1941 Professor Ernst Heinkel was developing the world's first jet-engined planes at the Heinkel plant in Rostock. In September 1943, the Me-262 jet fighter plane production – under the overall control of Field Marshal Erhard Milch, State Secretary in the Luftwaffe and armaments chief – was arbitrarily halted by Hitler.

Albert Speer wrote in his memoirs *Inside the Third Reich*:

The jet plane was not the only effective new weapon that could have been mass produced in 1944. We possessed a remote-controlled flying bomb, a rocket plane even faster than the jet plane, a rocket missile that homed on an enemy plane by tracking the heat rays from its motors, and a torpedo that reacted to sound and could thus pursue and hit a ship fleeing in a zigzag course. Development of a ground-to-air missile had been completed. The designer Lippisch had jet planes on the drawing board that were far in advance of anything so far known ... We were literally suffering from an excess of projects in development. Had we concentrated on only a few types we would surely have completed some of them sooner.

Hitler wanted the Me-262 converted into a light *bomber*.

In the summer of 1944, 8,000 V-1s exploded over London killing 5,000 civilians, wounding 40,000 and damaging 75,000 houses. The 'buzz-bomb' was a truly noisy weapon of casual destruction. This horrible campaign was carried out despite the SS arresting the Peenemünde scientists, including Werner von Braun, their chief designer. In spite of this the V-2, a more formidable weapon, was brought into action in September 1944. With a warhead of 1 tonne, it reached a height of 60 miles and plunged down on its target at a speed of over 3,500 mph. Its range was a little over 200 miles. Altogether about 6,000 were built with 1,265 launched against the port of Antwerp and 1,050 on London. The V-2 was 46 feet long and weighed more than 13 metric tonnes. Hitler wanted more than 900 produced monthly. Speer wrote: 'The whole notion was absurd. The fleets of enemy bombers in 1944 dropped an average of 2,000 tons of bombs *a day* over a span of several months. And Hitler wanted to retaliate with 30 rockets with 24 tons of explosives to England daily. That was equivalent to the bomb load of only 12 Flying Fortresses.' Speer admitted that he was at fault in supporting Hitler's decision:

That was probably one of my most serious mistakes. We would have done much better to focus our efforts on manufacturing a ground-to-air defensive rocket. Developed in 1942, code name Waterfall [translated], to such a point mass production would soon have been possible. The Waterfall was 25 feet long, carrying 660 pounds of explosives along a directional beam up to an altitude of 50,000 feet and hit the enemy bombers with great accuracy. It was not affected by day or night, by clouds or fog ... We could surely have produced several thousand of these smaller and less expensive rockets per month. To this day [1970] I think this rocket in conjunction with the jet fighters would have beaten back the western Allies' air offensive against our industry from the Spring of 1944 on.

As late as 1 January 1945, 2,210 scientists and engineers were working on the *long-range* rockets A-4 and A-9; only 220 were assigned to Waterfall, and 135 to another anti-aircraft rocket project, code name Typhoon.

The Mittelwerke facility, where the vengeance rockets were assembled underground with the use of slave labour.

In June 1944, the Germans unleashed their new V-1 (*Vergeltungswaffe* or vengeance weapon) pilotless flying bombs against targets in the south-east of England and Belgium. They were known popularly as 'buzz bombs' or 'doodlebugs' because of the charateristic noise made by their simple pulse jet engines.

Left: A captured V-2 rocket site, Leese, April 1945.
Right: A V-2 rocket being prepared for launch.

The A-4 had been trial-tested on 7 July 1943 and the film shown to Hitler. Speer wrote:

Hitler bade the Peenemünde men [Colonel Walter Dornberger and the 37-year-old von Braun] an exceedingly cordial goodbye. He was greatly impressed and his imagination had been kindled. Back in his bunker he became quite ecstatic about the possibilities of this project. 'The A-4 is a measure that can decide the war. And what encouragement to the home front when we attack the English with it! This is the decisive weapon of the war and what is more it can be produced with relatively small resources. Speer you must push the A-4 as hard as you can. Whatever labour and materials they need must be supplied instantly ... put the A-4 on a par with tank production. But in this project we can use only Germans. God help us if the enemy finds out about the business.'

166

From January 1945 Speer was confident that 210 Me-262s could be produced each month. A few were in action against the RAF and USAAF from 18 September 1944. In February 1945 production was 283. They needed exceptionally long runways, which were bombed heavily. They needed very experienced pilots and although the Luftwaffe were extremely proud of them, their overall effect was limited

There were two other forms of horror and vengeance weapons. Both Hitler and Churchill in 1939 decided to build limited stocks, but not to use them first. Corporal Hitler had been gassed in the First World War and knew what an unreliable weapon gas was, dependent entirely on the vagaries of the winds. The nuclear option is shown in an appendix.

So the elimination of the V-1, V-2 and other aerial deadly weapons was an important priority, even though the majority of sites had been captured or bombed. Large areas of northern Holland and north-west Germany could harbour all kinds of dangerous surprises.

Messerschmitt Me-262, Germany's jet fighter-bomber.

CHAPTER 11

Rhine Crossing Battle:
Operation Plunder

On 18 March General M. C. Dempsey, Commander of the 2nd British Army, briefed the key Allied journalists including Alan Moorehead at his camp near Venlo beside the river Maas: 'We will cross the Rhine on the night of the 23rd. Four corps are under my command. 30th Corps will cross in the north near Rees and continue northward for the capture of Emmerich. 12th Corps will cross in the centre at Xanten while the Commandos turn south to capture Wesel. On the following morning 18th US Airborne Corps with the 6th British Airborne Division under command will drop near the Ijssel River on the opposite bank of the Rhine and secure the bridges there. Finally the 8th Corps will follow through and I cannot yet tell you how far they will go. The 9th American Army will also attack on my right flank and proceed eastward along the Ruhr.' Alan Moorehead wrote:

Dempsey was an exact and energetic technician. Ever since the landing in Normandy, whenever we met him one was astonished at the precision with which he elucidated a plan and brought a map to life. Above and below him were two entirely different commanders; Montgomery with his flair, his processes of instinctive divination, his legendary influence with the generals and the soldiers. Then Horrocks of 30th Corps, a man with an ascetic, almost an ecclesiastical face and underneath that more volatility and fire and dash than one could have thought possible in a single man, an ideal break-through general. Both Montgomery (objectively) and Horrocks (subjectively) dealt often in terms of emotion. Between them stood Dempsey, the technician, a lean and nervous figure, a manipulator of facts, not so much a popular leader as a remarkable co-ordinator and a planner ... You must list him among the half-dozen really able field commanders thrown up by England in the war.

168

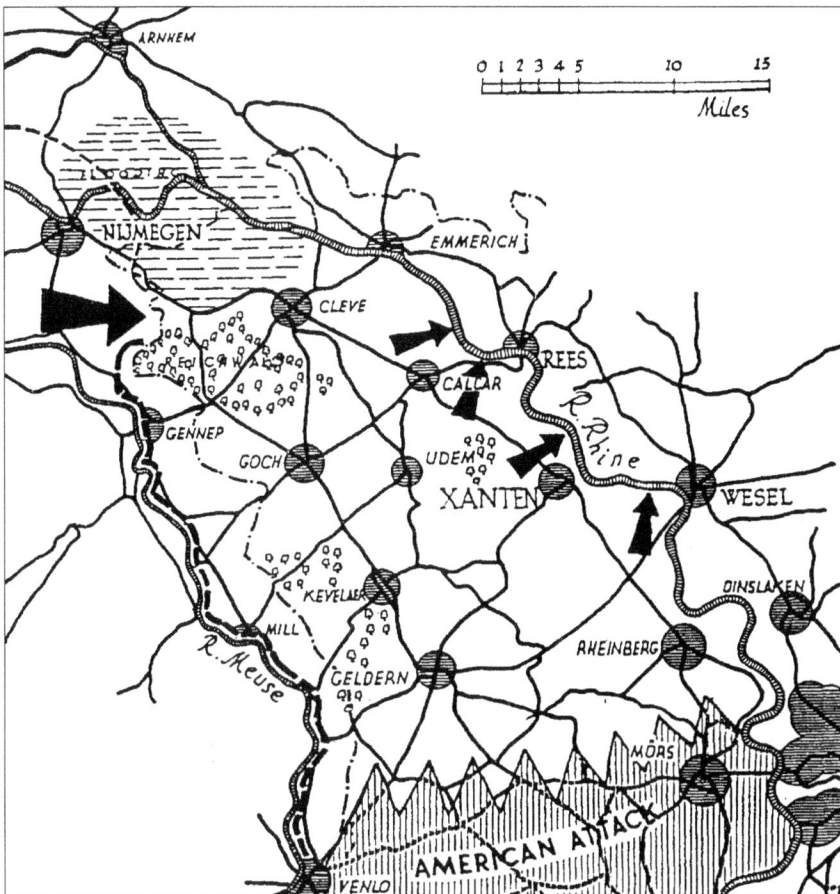

The battlegrounds of operations Veritable, Blockbuster and Plunder.

Operation Plunder was the code name for the assault crossing of the great river Rhine and for a deep thrust into the northern half of the Third Reich. Operation Varsity was the code name for the airborne drop by 6th Airborne and the US 17th Airborne Divisions, north and northwest of Wesel.

The RAF and USAAF had severely interdicted the German communications network in a line from Bremen southward to the Rhine and Coblenz. Most of the eighteen vital railroad bridges and viaducts were destroyed. In four weeks from 21 February, forty major bombing raids were carried out, including by the RAF dropping 22,000 lb monster bombs. On 11 and 12 March, over 1,000 bombers each night dropped over 5,000 tons on Essen and Dortmund. The new jet aircraft being introduced by the Luftwaffe had special attention and their jet airfields made unusable.

The Royal Navy carried out some strange manoeuvres. Large LCM and LCV(P) craft were transported overland and by waterway on special trailers up to 45 feet in length and 14 feet wide. Their ferry service would prove invaluable. An LCM could take a Sherman tank or sixty men, and a LCV (P) a bulldozer or thirty-five men. The width of the Rhine for Operation Plunder was between 400 and 500 yards, but at high water it could increase to between 700 and 1,200 yards. The current was usually about 3.5 knots (4 mph). The river bed was sand and gravel and amphibious Dual-Duplex tanks and trestles would have a good 'bearing surface'.

Supplies of equipment, guns, ammunition and smoke shells to shield the crossing poured into the west bank of the Rhine. Sappers had built three Class 70 and nine Class 4 bridges, together with railway bridges at Ravenstein and Mook, to allow the convoys of material to deploy. A railhead and store depot was opened near Goch. In all, 118,000 tons were needed, including 30,000 tons of engineering stores and bridging, which included 25,000 wooden pontoons, very vulnerable to shell-fire and ice, 2,000 assault boats, 650 storm boats and 120 river tugs, together with 80 miles of balloon cable and 260 miles of steel wire rope. Although this was primarily a Royal Engineers affair, it developed into a combined operation. For instance, the ferries and rafts had to be winched across by cable and there were not sufficient Royal Engineers for this purpose, as they were all required for other technical jobs, so 159th Wing of the RAF were asked whether they could spare some of the men who operated their balloons. They produced fifty specialists in 12 hours from a distance of 150 miles, and said that another 300 volunteers were available if required. This was typical of the assistance from the RAF. The Navy were equally helpful and produced a Royal Naval team which established anti-mine booms upstream to prevent the Germans floating down demolitions and destroying the bridges when they were constructed.

No fewer than 60,000 Royal Engineers and men of the Royal Navy were deployed. On the ground there were 400,000 vehicles of every possible description: tanks, SPs, AVREs, carriers, half-tracks, Crocodiles, Wasps, DUKWs, Crabs, DDs (swimming tanks), field ambulances, signals trucks, REME vehicles and more. An unusual group was the military

A smokescreen near the Rhine.

government detachments to administer occupied territory gained in Operation Plunder.

For ten days a smokescreen shrouded the Allied bank of the river opposite Wesel to conceal troop and vehicle movements from German artillery observers. 52nd Lowland Scottish Division was responsible for holding the western bank of the river with fourteen section posts between Buderich and Vynen, and their gunners knocked down the German OPs in the handsome churches in Wesel and Bislich. The ten-mile stretch ran in two wide loops and in the fields were many forlorn cattle, that needed milking, and poultry that laid eggs. In Xanten could be seen fleets of Buffaloes, DUKWs and DD tanks waiting for the off. So too were three eminent figures to be seen on the river banks 'consulting': a bulldog (W. Churchill), a bird (Montgomery) and a grave scholar in uniform (the CIGS, Alanbrooke)!

General Horrocks wrote:

> The 8th Parachute Division was in and around the town of Rees, with part of the 6th and 7th Parachute Divisions on their flanks, and supported by approximately 150 guns. Behind them, in reserve, were the 15th and 116th Panzer Divisions, all of whom we knew only too well from previous battles. Although they had

171

Above: Assault crossing of the Rhine.

Left: The Rhine at Rees.

Before dawn Scottish troops embark to cross the river Rhine. (*Imperial War Museum*)

suffered heavy casualties, these had been made up from within Germany. The bulk of their reinforcements of course were callow youths, since by this time the Germans really were scraping the bottom of the barrel to make up losses. But many of these youngsters were dedicated Nazis, and under the guidance of the extremely tough and experienced parachute and panzer officers and NCOs they had been soon moulded into a formidable fighting force. It was bad luck that once more we should be faced by these diehard Nazis; after the crossing we heard stories, from the US and our flanking corps, of German soldiers surrendering in their thousands, and of villages with white sheets hanging from every window, but this did not happen on our front. We had still to fight, and fight hard, right up to the end. I could not resist a feeling of admiration for the sterling qualities displayed by these German paratroopers and panzers, when the bulk of them must have known that the war was irretrievably lost.

Aerial photography had identified 357 German flak (anti-aircraft) positions with a total of about 1,000 guns, which would unfortunately prove extremely lethal.

Generals Dempsey and Horrocks' plan in Operation Widgeon was for the 1st Commando Brigade to cross the river at night to attack the

town of Wesel. Twenty-six Buffaloes of 77th RE Assault Squadron would
run a non-stop ferry service across. Each could hold twenty men with
equipment for the attack and thirty men for the follow-up. Field Marshal
Montgomery had specified two Scottish divisions to make the initial
break-in. 15th Scottish, in Operation Torchlight, were directed to attack
Bislich and Mehr, four miles west of Wesel. The two airborne divisions
would land in front around Hamminkeln and Diersfordter Wald, three
miles inland from Mehr and Bislich. One lesson had been learned from
the Arnhem battle: three miles probably okay, eight miles, disaster! The
51st Highland Division, part of 12th Corps under Lt General Neil Ritchie,
would cross on a two-brigade front, one near Honnopel and the other
downstream at Rees. Captain Angus Stewart, adjutant of the 7th Battalion
Argyll and Sutherland Highlanders, 51st Highland Division, described
being carried across the Rhine by the 1st Northants Yeomanry:

> At 0800 we got into our Buffaloes, ponderous great clumsy creatures looking
> very much like the original tanks of 1916. They stand about eight feet high on
> dry land and are driven both on land and water by great tracks which travel
> round the whole perimeter of the vehicle, instead of on bogies as on a modern

Operation Plunder, Rhine crossing, 24 March 1945. AVRE with fascine being ferried across
on a scissors bridge, supporting 15th Scottish Division. (*Birkin Haward*)

tank. These machines can carry a platoon of men or a small vehicle like a jeep
or a carrier.

At 0815 we started up, and there was a great din of engines revving and
guns firing. The 25-pdrs make a sharp crack, while the big heavy guns made a
resonant dong! Like the G string of a cello being plucked. What an orchestra!
I thought back to Medenine, Mareth, Enfidaville, Sicily and remembered all
those other times I had experienced this same experience. This time it was
rather different, though, more complicated, more hazardous, and just as vital.
This time our division assaulted alone; the next division on our flank didn't
cross till three in the morning, others not till ten, while the airborne divs were
also landing about ten. We were playing the good old game of trying to draw
his reserves onto us again, like Akele out hunting with pack.

At 0900 the first Buffaloes entered the water and chugged across to the
other side, none failed, none sank, all landed our men where we wanted them.
They were good, those crews. I followed about half an hour later with the
2 i/c and a carrier with a wireless set. Without difficulty we climbed out onto
the west bank of the Rhine after a fine crossing. I hardly knew when we were
in the water. We tilted our noses right up as we climbed over a bund and then
see-sawed on top with our noses right down. There was the water beneath
us, and we waddled unceremoniously into it. We looked quite graceful in the
water, low and manoeuvrable, though not at all fast. We had about six inches
freeboard and shipped nothing. When we came out on the far bank we found
a slight fire burning, so we put it out, let down our back and spewed out the
carrier. The companies by now were all on their objectives, the track was being
swept for mines and we started our job of fielding the vehicles as they came in
and disposing of them. So far, no enemy reply, and not a vehicle had been lost;
a remarkable feat.

The river crossing took about four minutes and on average it took
about 55 minutes to carry a battalion across the Rhine. The four Buffalo
regiments made 3,842 craft trips. Of the 425 Buffaloes operating only
nine were destroyed and fifty-five damaged, with casualties of only
forty.

There were soon efficient Class 50/60 rafting operations for light
vehicles, Weasels and anti-tank guns at crossings called Tilbury,
Gravesend, Ardath and Abdullah under an intense smokescreen and an
artillery barrage of 1,300 guns [the author's 13th RHA fired non-stop
for ten hours on selected fire programmes], with H-Hour at 2200hrs.

On the night of 23/24 March, Operation Plunder started under a clear full moon. The Commandos, then the Highland Division, and on the following morning 15th Scottish, crossed in Operation Torchlight, as described by Stewart Macpherson:

> I watched the Commandos take off for Wesel, long a thorn in our side, and their attitude simply defied description. You would have thought that they were embarking on a Union picnic: they just couldn't care less. A few minutes after they were due to arrive on the far side, Bomber Command were to deliver a crushing blow on the enemy in Wesel, while the Commandos lay doggo over there, a bare thousand yards from the RAF target, and waited. Bang on time Arthur Harris and Company [Air Chief Marshal Sir Arthur Harris was C-in-C Bomber Command] arrived and delivered a nerve shaking blow on the Wesel stronghold.

In pitch darkness with white tape guidelines, the Commandos, like thieves in the night, stole into the smashed up town and captured it quickly with the minimum of casualties. On the northern flank the 227th Brigade of the 15th Scottish, Gordons and Argyles leading, were tasked with making a bridgehead in the area Haffen-Mehr, which was held by the German 7th Parachute Division. Further south, 44th Brigade would cross from Xanten, led by the RSF and KOSB with the objective of Bislich, held by the battered 84th Wehrmacht Division. The assault crossing started at 0200hrs on 24 March under a 700-gun barrage and the divisional Pepperpot. This was a curious habit of the British Liberation Army on the doorstep of Germany. It needed careful co-ordination and was fired by supporting tanks, SPs, anti-tank guns, heavy mortars, LAA guns and medium machine-guns plus the Canadian Mattress rocket salvoes.

The whole operation went smoothly. The 1062nd Grenadier Regiment was overwhelmed; Bislich was captured, as were the villages of Vissel and Jockern. In mid-afternoon the KOSB and RSF met up with 17th US Airborne and the Royal Scots linked up with 6th Airborne. The only pitched battle was between a German parachute battalion and the HLI in the villages of Wolffskath, Bettenhof and Overkamp. The Scotsmen had 100 casualties in confused and bitter fighting. For the next few days the German 7th Parachute Division put up one hell of a fight defending vital bridges.

Assault infantry lands on the
east side of the Rhine.

The 51st Highland's assault crossing had gone to plan except for the
5/7th Gordons when their Buffaloes were beaten off by the Upper Rhine
defences, just east of Rees. 1st Gordons and 5th Black Watch suffered
badly in the five-day battle to capture Rees. Their splendid GOC, Major
General Thomas Rennie, was killed in a mortar stonk. General Horrocks
wrote: 'I have always felt that Rennie had some foreboding about this
battle. Like so many Highlanders, I believe he was "fey" ... All three
brigades were involved in heavy fighting so I crossed the Rhine in a
Buffalo and summoned the three brigadiers to a conference. They were
very upset ... no wonder because Rennie was a great leader.' The new
GOC was Major General 'Babe' Macmillan, commander of 49th Polar
Bears, who led 51st throughout Plunder and Eclipse to the end of the
war. General Horrocks wrote:

> In the darkness of that warm spring evening, I could imagine the leading
> Buffaloes carrying infantry of 153rd and 154th Infantry Brigades lumbering
> along their routes, which had been taped out and lit beforehand, and then
> lurching down into the dark waters of the Rhine.
>
> Upstream at Wesel I could hear the aircraft of Bomber Command preparing
> the way for 12th Corps which was to assault later that night. Then at four
> minutes past nine precisely I received the message for which I had been waiting
> – in its way a historical message because it was from the British troops to cross

A party of VIPs in front of the Citadel of Jülich, just after they had taken lunch in the ruined fortress. From left to right: Major General McLain (US 19th Corps), Field Marshal Montgomery (British 21st Army Group), Winston Churchill (British Prime Minister), Major General Gillem (US 13th Corps), Field Marshal Allan Brooke (British Chief Imperial General Staff) and Lt General Simpson (US 9th Army). (*Imperial War Museum*)

the Rhine – 'The Black Watch has landed safely on the far bank.' The initial crossings went very smoothly, opposition was not as heavy as might have been expected and our casualties were comparatively light.

But the enemy was quick to recover and very soon reports came in that Rees was proving troublesome. As the night wore on enemy resistance stiffened and some very bitter fighting took place. Within 24 hours of the assault the 15th Panzergrenadier Division hit back at us with a vicious counter-attack ...

Reports were coming in of Germans surrendering in large numbers to the British and American forces on our flanks but there was no sign of any collapse on our front. In fact the 51st Highland Division reported that the enemy was fighting harder than at any time since Normandy. It says a lot for the morale of those German parachute and panzer troops that with chaos, disorganisation and disillusionment all round them they should still be resisting so stubbornly. Their casualties during the last nine months had been very heavy, and the reinforcements arriving from Germany had not been of the old calibre at all, yet somehow the rough experienced officers and NCOs who were such a feature of these parachute and panzer formations managed to turn the callow youths into good soldiers.

It was a slow business widening the bridgehead and Rees proved a particularly hard nut to crack. It took the 1st Battalion, the Gordons, 48 hours of dour fighting before the whole place was in their hands.

Robert Barr, 52nd Lowland, noted that the 'bulldog' and his companions had come to see his warriors on the Third Reich battlefields:

25 March 1945. In warm, brilliant sunshine this afternoon the Prime Minister, Mr Churchill, basked on his balcony overlooking the Rhine and discussed casually with General Eisenhower and Field Marshal Montgomery just how the 9th Army bridgehead had been established. Downstream he could see the town of Wesel which the British Commandos had just completely cleared. Through his binoculars the Prime Minister was inspecting the bridgehead just across the slow-flowing Rhine, when quite suddenly he decided to cross. Our planes were still pounding the German positions across the river when Mr

Churchill in a jeep with Monty and Dempsey on the banks of the Rhine.

Churchill aboard a Daimler armoured car Xanten/Calcar, before the Rhine crossing, 25 March 1945.

Churchill walked down to the river's edge and got into a landing-craft. With him went Sir Alan Brooke, Chief of the Imperial General Staff, Field Marshal Montgomery, General Omar Bradley, and General Simpson, Commander of the 9th US Army which had forced the river at this point. We cruised across the Rhine in the tracks of the infantry, and the Prime Minister scrambled up the gravel bank on the other side along the same narrow wired path that the infantry had used, and scaled a high earth dike to get a good view. He studied the bridgehead and discussed the morning's battle with the American generals and then he decided to have a short cruise on the Rhine. It was his first cruise on the Rhine since he sailed it in a motor torpedo boat at the end of the last war. After the short cruise, the landing-craft turned back to shore. The Prime Minister went to Wesel along the river road. At the approach to the great steel and concrete Wesel bridge (the one the Germans destroyed when they retreated across the river) the cavalcade of staff cars halted and the Prime Minister got out. Leaving the party, he found a path through the debris, and climbed up on to the first span of the wrecked bridge. The ruins of Wesel were just across the water, and the sound of machine-gun and rifle fire was still coming up spasmodically from the town. Through binoculars the Prime Minister watched

Churchill crossing the Rhine.

the crouching infantry moving through the streets, while Field Marshal Montgomery explained how the British assault had been made.

Alan Moorehead noted:

… naval launches flying the ensign were plying to and fro. Already the bridge builders were at work. The Buffaloes lay about in the fields in the sunshine like prehistoric monsters, their exhausted crews spreadeagled out on the grass asleep after the night's work … Long lines of prisoners go by carrying and supporting their wounded. An enterprising Military Government officer was distributing General Eisenhower's proclamation in English and German in the [captured] villages. Nazi emblems must be destroyed. All Nazi organisations and courts of law are abolished. Householders must place lists of names of the inmates on the front doors of their houses.

CHAPTER 12

Airborne Attack: Operation Varsity

This was the last great airborne operation of the Second World War. Much thought had gone into the planning for the 1st Allied Airborne Army, so that the painful experience of Market Garden would not be repeated. First of all the whole assault would be made in one lift, not scattered over three days, secondly the whole force would land, not near, but on top of its objective. There would be no long approach march. To ensure this, the operation would take place in broad daylight – a great but necessary risk. Finally the airborne troops would land *after* the large-scale land operations had crossed the Rhine. Only when the key German town of Wesel had been captured by the Commando Brigade (with total casualties of thirty-six!) would the airborne armada arrive.

The 17th US Airborne Division had not been in action before and 6th Airborne not since their heroic battles in Normandy and the Ardennes.

The objectives were for 3rd and 5th Parachute Brigades and the Airlanding Brigade to occupy the high ground east of Bergen in the wooded area of the Diers-Ijssel fordter Wald, the village of Hamminkeln and certain bridges over the river Ijssel.

After Market Garden, Lt General Browning had been 'posted away' and the experienced Major General R. N. 'Windy' Gale took his place as Deputy Commander of 1st AAA. Major General E. Bols had commanded 6th Airborne in the Ardennes and did so for the rest of the war.

The 6th Airborne flew from bases in East Anglia in 669 planes and 429 gliders of the RAF 38th and 46th Groups, while the 17th US Airborne flew from the Paris area in 903 planes and 897 gliders.

The British landed 7,220 officers and men between 1000 and 1300hrs on 24 March. Fighter escorts of 213 RAF and 676/9th USAF

Horsa gliders soar over the Rhine despite smoke and haze. (*Imperial War Museum*)

6th Airlanding Brigade land near Hamminkeln, 27 March 1945.

protected the two armadas. General Eisenhower noted: '46 planes destroyed (3.98% of those employed) was remarkably low … since no evasive action was taken.' Nevertheless British casualties were 1,400 (347 killed in action) and Americans 1,500. 60 per cent of the British gliders and 50 per cent of the American gliders suffered flak damage. Twenty-two of the seventy-two C46 transports were lost and General Ridgeway was so dismayed by the ease with which they had caught fire, he forbade their use for carrying paratroops in the future.

6th Airborne gliders lifted 4,844 men, 342 jeeps, 348 trailers, 7 Locust tanks, 14 lorries, 2 bulldozers, 11 Bren-carriers, 19 Scout cars, 215 motorcycles, 68 assorted guns, mainly anti-tank, and 10 mortars.

The journalist Richard Dimbleby flew with 6th Airborne and reported what he saw:

24 March 1945. The Rhine lies left and right across our path below us, shining in the sunlight – wide and with sweeping curves; and the whole of this mighty airborne army is now crossing and filling the whole sky. We haven't come as far as this without some loss; on our right-hand side a Dakota has just gone down in flames. We watched it go to the ground, and I've just seen the parachutes of it blossoming and floating down towards the river. Above us and below us, collecting close round us now, are the tugs as they take their gliders in. Down there is the smoke of battle. There is the smoke-screen laid by the army lying right across the far bank of the river; dense clouds of brown and grey smoke coming up. And now our skipper's talking to the glider pilot and warning him that we're nearly there, preparing to cast him off. Ahead of us, another pillar of black smoke marks the spot where an aircraft has gone down, and – yet another one; it's a Stirling – a British Stirling; it's going down with flames coming out from under its belly – four parachutes have come out of the Stirling; it goes on its way to the ground. We haven't got time to watch it further because we're coming up now to the exact chosen landing-ground where our airborne forces have to be put down; and no matter what the opposition may be, we have got to keep straight on, dead on the exact position. There's only a minute or two to go, we cross the Rhine – we're on the east bank of the river. We're passing now over the army smoke-cloud. 'Stand by and I'll tell you when to jump off.' The pilot is calling – warning us – in just one moment we shall have let go. All over the sky ahead of us – here comes the voice – *Now!* – The glider has gone: we've cast off our glider.

6th Airborne troops
after capturing Wesel.

We've let her go. There she goes down behind us. We've turned hard away,
hard away in a tight circle to port to get out of this area. I'm sorry if I'm
shouting – this is a very tremendous sight.

The ground fighting went according to plan; most of the casualties
were during the drops and around the dropping zones. Corporal F. G.
Topham, a medical orderly with 1st Canadian Parachute Battalion, over
a six-hour period aided many wounded men, though wounded himself,
and was awarded the Victoria Cross. Stanley Maxted was there and
reported:

Down toward the river beyond the trees a company of men were deployed. The
glider-borne troops involved in this part of the action wear the badges of three
very famous old county regiments – the Royal Ulster Rifles, the Oxfordshire
and Buckinghamshire Light Infantry, and the Devons. In some other setting
their actions might have reminded you of some nursery game. One section
would rise up and go forward like Indian bush fighters. They'd be covered by
fire from the rest. Then another section went forward in the same way until all
the platoons had worked their way in a converging movement on to the bridge
that was one of their objectives.

It really is an occasion, to watch these airborne troops fight. When they first
come out of their gliders, with magazines blazing, to fight for their landing

Detail of picture on page 121.

zone it's more or less or a scramble, but from then on their operations are cold – deadly methodical.

On 24 March Churchill stood on Xanten hilltop and looked down across the morning battle mist at the place where the troops were still crossing on boats and rafts. 'I should have liked to have deployed my men in red coats on the plain down there and ordered them to charge.' And he added with vim, 'But now my armies are too vast.' Alan Moorehead watched Plunder and Varsity:

The first airborne troops ... a wonderful sight. They passed over only two or three hundreds of feet above our heads, the tow planes drawing sometimes one and sometimes two gliders and flying in tight formation. Then single planes with the parachutists waiting intensely inside for the moment to plunge through the open hatches. Here and there among all these hundreds of aircraft one would be hit by ack-ack fire and it was an agonising thing to see it break formation and start questing vainly back and forth in search of any sort of landing field

186

and then at last plunge headlong to the ground. Within a few minutes nothing would be left but the black pillar of petrol smoke and the unidentifiable scraps of wings and propellers and human beings ... The Germans were waiting for the landing mainly because there was only one obvious place to drop in the vicinity and this was it. A heavy mist, possibly created by the enemy, covered the dropping zone. Unable to see through it at the last moment, in the face of flying shrapnel many pilots judged the ground to be either higher or lower than it was. And so many crashed into farmhouses and ditches. Others broke up in the air and others again dropped sheer to the earth from 30 or 40 feet. At the same time the falling parachutists were picked off in the air during that vital few minutes. In the end the British Airborne Division had nearly 40 percent casualties. But this was to be no second Arnhem ...

Wynford Vaughan-Thomas witnessed the linkup between 6th Airborne and the Scotsmen:

... once on the other bank you do get a feeling of being well over this last barrier before the heart of Germany. For this bridgehead isn't a mere toe-hold on the other bank; you can now drive well inland, past the riverside farms, most of them pretty smashed up by our artillery fire and all with white flags

German prisoners captured on the east bank of the Rhine.

187

BLA armour link up with Airborne troops. Gliders and German POWs can be seen in the background. (*Imperial War Museum*)

hanging out of the few remaining windows. And then up into the belt of woodland, where we first contacted the parachute division. The wood was full of our parachutists, digging in, and beyond the wood that fantastic coloured landscape you always get after a big airborne landing, with parachutes dangling from the trees, green parachutes, yellow parachutes, and containers scattered over the fields, gliders flung broadcast over the ploughed land. It looked as if a gigantic litter basket had been emptied over the whole countryside. And, in the middle of the litter, small groups of cowed and completely bewildered civilians were gathered at nearly every farmhouse. They were listening patiently while they were lectured by a paratrooper on how they were to behave in future. They were giving no trouble, there were no desperate guerrilla fighters in this lot, they were only too eager to obey. And they didn't even look up as the long lines of tanks and guns came through the woods to join the paratroops. For in this section our link-up with the airborne landing is no narrow corridor, it's complete and the front is united. And all the way in from the river-bank the roads are full of new troops coming up, and infantry marching through the dust, and the farmhouses being taken over as headquarters, ration dumps being set up in the fields. All the signs of the army moving in, in a big way.

Chester Wilmot wrote about the amazing 'Red Berets' at the end of Operation Varsity:

The Airborne paratroops, lacking transport, 'scrounged' everything on wheels – from prams to steam engines!

29 March 1945. They're trained for this kind of footloose fighting on a fluid front, and they thrive on it. At one point yesterday afternoon they struck quite a strong German position blocking the road, and three tanks of the Airborne Reconnaissance Regiment were soon knocked out. Undismayed, the paratroops worked round through the woods, waited for dusk, and then stormed the guns that had held them up. They captured two 75s and four light ack-ack guns and their crews.

Yesterday and last night, as they beat through the woods, they came upon half a dozen self-propelled guns. They ambushed two of them and knocked them out with anti-tank bombs. They captured four others intact, two of these with the crews still sleeping inside.

I saw some of their captured armour on the road today. They had pressed it into service and had found scratch crews from among their own paratroops. The SP guns had joined the strange convoy of airborne forces on the road eastwards today. The airborne troops are always short of transport. They can't bring much with them, and they rely on capturing vehicles from the enemy. And now the 6th Airborne has a fine assortment. As I drove along, I passed men on German bicycles, in Volkswagens, in farm-carts drawn by draught horses. One man had a smart jinker, and another was driving a German bulldozer towing a German trailer. Some of the troops were carrying their ammunition and packs in little handcarts they had found on farms, and some were even pushing their equipment in perambulators.

189

CHAPTER 13

Affairs of State

At the Yalta Conference, 4–10 February 1945, Stalin and Roosevelt had marginalised Churchill, as they had done previously at Tehran. One key subject was how the conquering Allies should treat the Third Reich post-war. Stalin and Roosevelt exchanged 'brutalities', wanting a very severe settlement for the post-war Germany. Churchill remembered that the Versailles Treaty at the end of the First World War had totally humiliated Germany and thus produced a political climate in which Hitler quickly and shrewdly took great advantage. This was the official statement released to the press:

> It is our inflexible purpose to destroy German militarism and Nazism and to ensure that Germany will never again be able to disturb the peace of the world. We are determined to disarm and disband all German armed forces; break up for all time the German General Staff that has repeatedly contrived the resurgence of German militarism; remove or destroy all German military equipment; eliminate or control all German industry that could be used for military production; bring all war criminals to just and swift punishment and exact reparation in kind for the destruction wrought by the Germans; wipe out the Nazi party, Nazi laws, organizations and institutions, remove all Nazi and militarist influences from public office and from the cultural and economic life of the German people; and take in harmony such other measures in Germany as may be necessary to the future peace and safety of the world. It is not our purpose to destroy the people of Germany, but only when Nazism and Militarism have been extirpated will there be hope for a decent life for Germans, and a place for them in the comity of nations.

On 1 April 1945 it was possible to feel sorry for Prime Minister Winston Churchill. At the Yalta Conference in the Crimea with Stalin, Molotov, Gromyko, President Roosevelt and assorted military and political grandees (twenty-three altogether) from the three great powers, the future of the Western world was, more or less, agreed. There were three interpreters at this historic meeting; one of them was Pavlov, which may explain what eventually happened! Churchill's speech at the Yusupov Palace included: 'We regard Marshal Stalin's life as most precious to the hopes and hearts of all of us. There have been many conquerors in history, but few of them have been statesmen and most of them threw away the fruits of victory in the troubles which followed their wars. I earnestly hope that the Marshal may be spared to the people of the Soviet Union ... I walk through this world with greater courage and hope when I find myself in a relation of friendship and intimacy with this great man whose fame has gone out not only over all Russia but the world.'

Operation Argonaut was the code name of this historic meeting, or as Churchill dubbed it, 'the Riviera of Hades'. The future of Europe, including Poland, was discussed and agreed – except that when Stalin agreed to anything it meant: 'I hear what you say and have no intention of implementing these promises.' Quite soon there were indications that 'Uncle Joe' did not mean to honour any Yalta agreements at all.

President Roosevelt was sick and ailing and died on 12 April, aged sixty-two. Churchill wrote: 'When I received these tidings ... I felt as if I had been struck a physical blow. My relations with this shining personality had played so large a part in the long terrible years we had worked together. Now they had come to an end and I was overpowered by a sense of deep and irreparable loss.' Hitler of course was delighted and hoped that Roosevelt's successor would concentrate on the war with Japan.

Then suddenly, out of the blue, General Eisenhower, without a word to the combined Chiefs of Staff, to his deputy, Air Chief Marshal Tedder, or indeed to Churchill, on 28 March sent a telegram to Marshal Stalin, which tacitly meant that the western Allies would not attempt to capture Berlin. Churchill had always advocated capture. Now Stalin readily agreed, saying that the proposal 'entirely coincides with the plan of the Soviet High Command. Berlin has lost its former strategic importance.' Stalin, of course, was lying through his teeth. It was vital to the prestige

Field Marshal Montgomery and General Eisenhower confer.

and honour of Russia to capture and probably destroy the arch enemy who had launched Barbarossa against it, had probed the defences of Moscow and inflicted millions of casualties – Berlin had to be taken. Revenge is sweet.

Churchill was irked that Eisenhower had acted so arbitrarily and unilaterally, but Ike had taken soundings from his top-ranking generals and the consensus was that the rendition of Berlin would cost at least 100,000 American casualties plus other Allied forces. Politically this would be totally unacceptable. There were no votes back at home for capturing the capital of the Third Reich. He wrote:

> Berlin, I was now certain, no longer represented a military objective of major importance. The Russian advance and the Allied bombing had largely destroyed its usefulness ... military factors when the enemy was on the brink of final defeat was more important in my eyes than the political considerations involved in an Allied capture of the capital. The function of our forces must be to crush the German armies rather than to dissipate our own strength in the occupation of empty and ruined cities.

MONTGOMERY'S PLAN

Hanover

THE RUHR

Kassel

Calais

Scheldt

Antwerp

Brussels

Cologne

Coblenz

Frankfurt

ARDENNES

Moselle

Rhine

Cherbourg

Dieppe

Le Havre

Caen

Rheims

THE SAAR

Mannheim

Marne

Meuse

Metz

Karlsruhe

FIRST CON ARMY

SECOND BR ARMY

FIRST U.S. ARMY

THIRD U.S. ARMY

PARIS

Nancy

Strasbourg

Seine

Orleans

Loire

Dijon

Siegfried Line

SCALE OF MILES

0   50   100   150

SEVENTH U.S. ARMY
(DIVERSIONARY THREAT ONLY)

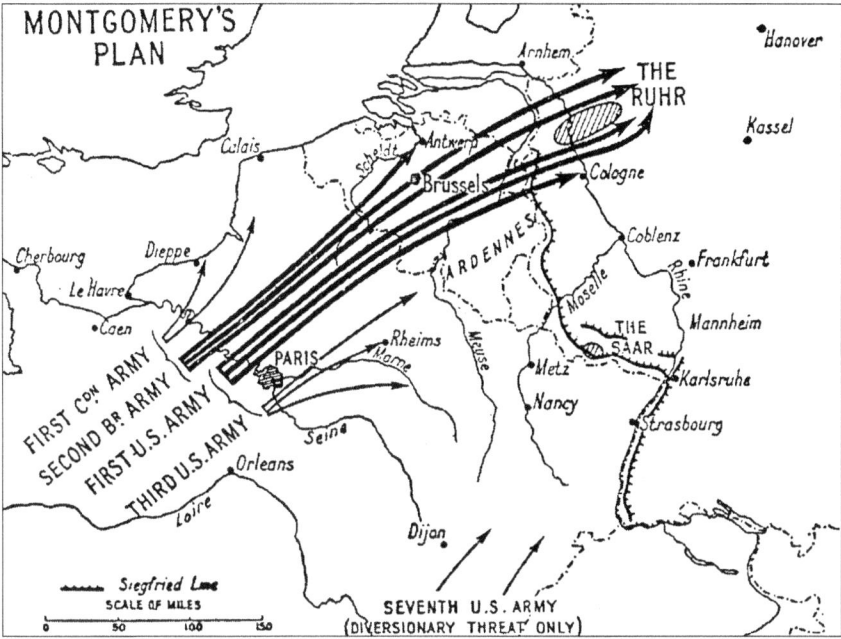

Above and below: Montgomery's and Eisenhower's plans differ!

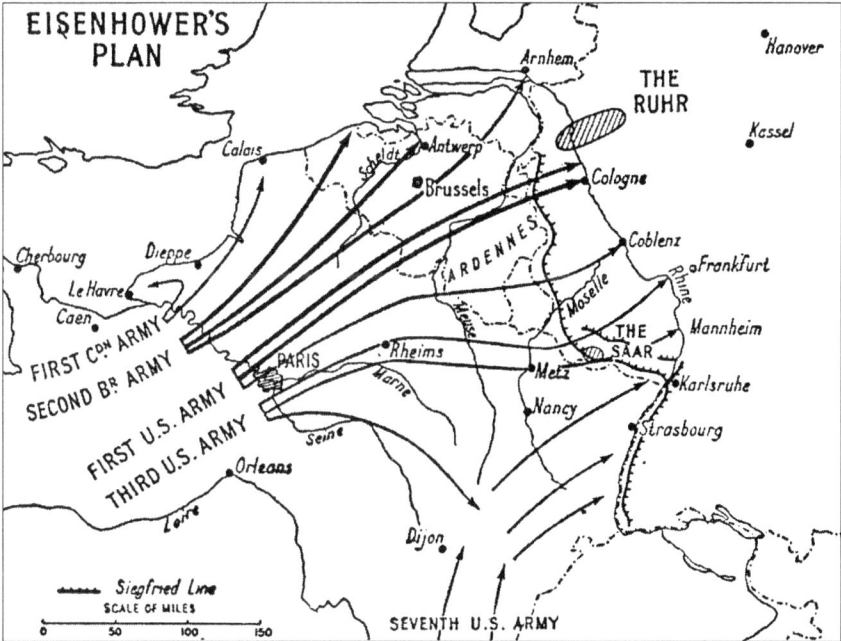

EISENHOWER'S PLAN

Hanover

Arnhem

THE RUHR

Kassel

Calais

Scheldt

Antwerp

Brussels

Cologne

Coblenz

Frankfurt

ARDENNES

Moselle

Rhine

Cherbourg

Dieppe

Le Havre

Caen

Rheims

THE SAAR

Mannheim

Meuse

Metz

Karlsruhe

FIRST CON ARMY

SECOND BR ARMY

FIRST U.S. ARMY

THIRD U.S. ARMY

PARIS

Marne

Nancy

Strasbourg

Seine

Orleans

Loire

Dijon

SEVENTH U.S. ARMY

Siegfried Line

SCALE OF MILES

0   50   100   150

193

Churchill and Montgomery were totally wrong even to contemplate the siege and capture of Berlin. Montgomery's forces were dwindling each day; 5th Infantry Division was summoned from Italy; the Canadians imported massive reinforcements, also from Italy. It would have been lunacy for the battered British Liberation Army to lay siege to Berlin. Marshal Zhukov's huge armies had immense losses in their seizure of Berlin.

The balance of power between the Allies had been swinging powerfully away from Churchill and his British armed forces almost since D-Day. Eisenhower now had under his command an overwhelming majority of forces: over two-thirds of the fighting men on the Western Front. Churchill's influence was waning, week by week, and after five years at the helm that was a difficult situation to accept.

Churchill soon realised that Stalin did not mean to keep to the policies agreed at Yalta. Poland had had a tragic history during the war, over-run in a few weeks by Hitler's panzers in 1939/40, ten thousand Polish officers murdered at Katyn (known to Churchill through intelligence from Ultra) and now it was obvious that Stalin's Red Army would occupy the country as the Germans were driven out. It was obvious too that Stalin was in effect conquering the Balkans and almost certainly had designs on the Baltic States, including Denmark.

With President Roosevelt at death's door, April 1945 was indeed a very difficult month for Churchill, even as his legions battled their way across the Rhine.

CHAPTER 14

Over the Rhine

Max Hastings wrote in his book *Armageddon*: 'Montgomery's Rhine operation was plodding and over-insured, but those [including this author] who crossed the water would be grateful that their objectives were gained at small price. But the casualties incurred by the airborne assault were out of all proportion to its contribution. Gliders were never again employed in war. Operation Varsity was a folly for which more than a thousand men paid with their lives – almost as many as 1st Airborne lost killed at Arnhem. Once again, a baleful reality had been permitted to steer events.'

Certainly most of the American generals mocked Montgomery's ultra-careful methodical build-up for Operation Plunder. After all, Lt Karl Timmerman, with the US 27th Armoured Infantry, had brilliantly and bravely captured the bridge at Remagen and helped set up a bridgehead over the Rhine – all in a matter of hours with few casualties.

In Max Hastings' view, Varsity was a 'spectacular shambles', but he is not correct. Montgomery was a careful commander, apart from the dash towards Arnhem. All his operations had been the same: a set-piece plan involving RAF saturation bombing, a massive artillery bombardment, and then the unleashing of his infantry and armour. In many cases this was successful, in many cases it was an attritional 'draw'. The attempt at Arnhem to capture a lodgement over the Rhine was unsuccessful, with sadly heavy losses. Operation Plunder was 100 per cent successful: the Rhine was crossed easily and quite soon his 'dogs of war', 11th Armoured, Desert Rats, the Guards and Polish armour were 'swanning' through Germany.

In operations Veritable and Blockbuster, in the dreadful mire of the Reichwald and Hochwald, the German defenders were, unfortunately,

6 KOSB over the Rhine, near Bislich.

magnificent. Lt Edwin Bramall (well before he became a field marshal) wrote from the perspective of a KRRC platoon commander at the 'sharp end':

> The Reichswald was the nastiest battle we had fought since Normandy. The Germans had constructed five successive lines of defence, manned chiefly by paratroopers. Flooded ground on both flanks forced the British and Canadians to advance on a narrow front. The thick woodland was almost impenetrable to tanks. Foliage jammed [tank] turret traverses. The Shermans were unable to use their armament effectively – it was too dangerous to fire high-explosive shells lest they hit trees above their own infantry. The tanks were also highly vulnerable to 'faust' ambushes in the dense cover.

General Schlemm had fought a masterly retreat with his paratroop divisions, the *crème de la crème*, fighting every inch of the way back to Xanten and Wesel. Undoubtedly the German formations facing the British and Canadians in the north of the Siegfried Line defences were more resolute and experienced than those in the south. All the combined

A Squadron 3 RTR in Operation Plunder. (*Tank Museum*)

intelligence sources at SHAEF had been totally wrong in the autumn of 1944 in predicting the end of the war 'by Christmas' with their schoolboy cries of 'On to Berlin'. Field Marshal Montgomery could not be sure that his troops would not encounter the same expert and determined defences across the Rhine. He was taking out an insurance policy that the British Liberation Army would get across the Rhine, this time, with the minimum of casualties. His foot soldiers were now a shadow of the fine vigorous 'virgin' divisions which had landed on D-Day nine or ten months back.

The month in the Reichswald had caused considerable casualties. It had proved that the anti-tank regiment and the light anti-aircraft regiment in *every* British division, eleven of them in all, were probably not needed in Operation Eclipse. German panzers were now few and far between and the Luftwaffe was rarely seen. In the armoured divisions the anti-tank regiment was transformed into an extra armoured infantry unit and was greatly appreciated as such.

Edgar Palamountain, staff officer of 11th Armoured Division, in his history *Taurus Pursuant*, wrote a description of the aftermath of Operation Plunder:

The word 'irruption' conveys a force invading a region [the Third Reich] with speed and violence and making a deep penetration into it. Neither 'invasion' nor 'penetration' would sufficiently indicate the pace at which this process was carried out. 'Breakthrough' supplies this deficiency; but 'breakthrough' has come to mean a violent penetration of a defensive position followed by a period of plain sailing which is, strictly speaking, the exploitation of the breakthrough, and this interpretation had been strengthened by certain well-remembered aspects of the campaign in France. In this sense the advance inside Germany was certainly not a breakthrough, for its momentum was interrupted not merely by obstacles and demolitions, but by enemy forces ready to fight hard battles in defence of the positions which they occupied. If there is any other phase of the total campaign which the advance in Germany may be said to resemble, this is probably the offensive through the bocage from Caumont to Laigle [Operation Bluecoat]. There are, however, certain obvious differences, the chief of which is that the Germans were now operating so far inside their own country that the execution of planned withdrawals must have seemed to them somewhat of a mockery. When you are fighting on your own soil you must sooner or later reach a stage where every retreat is a defeat.

Royal Scots Fusiliers move through woods over the Rhine under shellfire.

To describe the condition to which the German armies and the German State were now being reduced by the various invasions, the word 'disruption' could be used. But its employment would remove the spotlight from the division whose story this is and focus it rather on the enemy in his disintegration. Or, alternatively, it might suggest that that disintegration was a direct result of the activities of 11th Armoured; and this would be presumption indeed. Germany lay pierced by many wounds; only one of these was a cornada, a gash inflicted by the horn of the bull. [The black bull was the 11th Armoured insignia]

Montgomery and Dempsey carefully planned the continuation of Operation Plunder into the heart of Germany. The US 9th Army, under Monty's command, on the right, southern flank, then 8th, 12th, 30th Corps, and in the north the 2nd Canadian Corps. All would be led by an armoured division plus the British 4th (Black Rat), 8th (Red Fox's mask) and 6th Guards independent armoured brigades. The river Ems and the Dortmund-Ems Canal were the next hurdles to be encountered.

Lt General Horrocks wrote about his 'Corps de Chasse', armoured divisions available to him in 30th Corps:

The two operated quite differently. The 11th Armoured was the more flexible of the two. Pip Roberts, one of the most experienced tank commanders in the Army would switch his troops rapidly, attacking with tanks only, tanks and infantry, or infantry supported by tanks according to the situation. The Guards on the other hand always worked in regimental groups, Coldstream tanks and Coldstream infantry, Grenadier infantry and Grenadier tanks, etc. The system was not calculated to get the full value from the mobility of an armoured division, but it ensured the closest possible co-operation between the two arms, as the infantry and tank crews all came from the same regiment and had more or less grown up together. Nevertheless I preferred the standard armoured divisional practice of the 11th Armoured Division. Fortunately by now I had gained sufficient battle experience not to interfere with the way a divisional commander handled his command.

In Market Garden the Guards were ponderous, took their time and were not flexible enough, which contributed to the lack of success of the operation. Now in the wide open spaces of Westphalia the Guards did well against their old enemies, 15th Panzergrenadier Division and the 1st Parachute Army. Operation Plunder set off the extraordinary armoured plunge into the very guts of the Third Reich, which was dying and doomed. Everybody in the world knew it: Russians, Americans, British, Canadians, French and Polish formations knew it – everybody except the Austrian corporal in his Berlin bunker. Many of his legions, certainly the SS, the skilled parachute divisions and the renegade SS [Dutch, Vikings from Scandinavia and the East/Ost] were prepared to fight until the last possible moment. The horrors of death came suddenly to the advancing Allies in Operation Eclipse, in many ways. The Hitler Jugend and indeed the young schoolboys equipped with panzerfaust could and did destroy gleaming new British Comet tanks. The unemployed marine divisions buried immense 500-pound sea mines around Bremen and Bremenhaven, which disintegrated lead formations. What lay ahead of the British Liberation Army in their corridor – about 100 miles wide and 250 miles in length on the way to the Baltic – was of course not known. But the astonishing fight back by the Führer's forces in Market Garden, the Ardennes, and operations Blackcock, Veritable and Blockbuster surely could not continue – or could it? Five more river barriers and the two great cities of Bremen and Hamburg lay ahead.

Were there more secret weapons ahead – maybe vengeance weapons, perhaps gas, or chemical, even atomic (although that concept was not known outside the 'Tube Alloys' cognoscenti at the time)? 'Over the Rhine' was an extraordinary, daunting prospect. 'Bomber' Harris had destroyed every major city and town in Germany, but in Normandy, Holland and in the western Rhineland, demolished towns were defended tooth and nail by German kampfgruppen. Would that continue inside the Third Reich?

Street fighting was hated and smashed towns and villages favoured the defenders. Among houses, tactical radios, usually unreliable 19- and 21-sets, all ceased to function. Tanks were vulnerable to petrol bombs, grenades, and especially to the cheap one-shot panzerfausts. 'Clearing a town is an arduous process which cannot be hurried,' was a British briefing dogma.

So the leading reconnaissance units – the Cherry Pickers, the Inns of Court, the Household Cavalry and others – would encounter a small village, perhaps with obstacles and felled trees blocking the road. The roadside verges might be mined. The lead vehicle and its crew were always at risk and with the war inevitably, slowly, coming to an end, the easiest military tactic was to put an artillery stonk into that hapless village, just in case. And that happened frequently.

200

On 30 March, Stadtlohn, north-east of Wesel had been heavily bombed by the RAF. The Royal Engineers built this Bailey bridge over the river.

Another alternative was to introduce a Crocodile troop if the village was known to be held by the enemy. Lt Andrew Wilson commanded one of these terrible Churchill flame-thrower troops of the Buffs and wrote in his book *Flamethrower*:

He looked for places where the enemy might still be hiding. There was a wooden shed. As the flame hit it, the wood blew away in a burning mass, and there in the wreckage was the body of a Spandau [highly effective machine gun] … The [tank] gunner gripped his trigger. Swinging the turret down the front of the burning village, he began firing off the 75 [tank gun] at point-blank range. Where, oh where was the [supporting] infantry? As always in action he lost count of time. Wherever he swung the cupola, he saw fire and smoke and the track of destruction. Then all at once it was over. By the barn a little group of grey-clad Germans appeared without helmets or weapons, waving a sheet or a pole. He gave the order to stop firing and opened the hatches. The air was full of smuts and the sickly-sweet smell of fuel. Eventually there were 30 or 40. They moved slowly. In the hush of the moment, he felt a great elation; if ordered he could have driven through the smoking village right on to the enemy's divisional HQ. Nothing could have stopped him. He couldn't be harmed. Then the infantry came swarming into the village, dodging the mortar shells which the enemy had started dropping down. In the confusion the Germans began to bring out their wounded, blinded and burned, roughly bandaged beneath their charred uniforms. Some of them looked at the Crocodile. What were they thinking? He went back to refuel.

CHAPTER 15

Battle for the Teutoburger Wald

Hitler had decided to replace the veteran Field Marshal von Rundstedt, and selected Field Marshal Albert Kesselring to defend the right bank of the river Rhine against Eisenhower's huge armies of nearly four million men. Kesselring had performed defensive miracles on the Italian front behind the Hitler and Gothic Lines. He was a stubborn and resolute man with a blunt jaw and stocky figure and Hitler expected miracles from him. General Günther Blumentritt took over command of the 1st Parachute Army on 27 March as General Schlemm had been wounded during Operation Blockbuster. Blumentritt wrote:

> I reported to Colonel General Blaskowitz at Army Group H and he and his Chief-of-Staff both agreed that once the Rhine had been crossed the situation, with the forces at our disposal, was past repair. But since orders from above were to continue resisting I was to do the best that I could. When I took over my new command on 28 March I found that there were great gaps in my front, that I had no reserves, that my artillery was weak, that I had no air support whatever and hardly any tanks. My communication and signal facilities were entirely inadequate and there was one corps under my command that I was never able to contact. The reinforcements that still came to me were hastily trained and badly equipped, and I never used them so that I could save needless casualties.
>
> Nevertheless orders from the Supreme Command were still couched in the most rigorous terms, enjoining us to 'hold' and 'fight' under threats of court-martial. But I no longer insisted on these orders being carried out. It was a nerve-racking time we experienced – outwardly putting a bold face on the matter in order to do one's duty as one had sworn to do – while we secretly

allowed things to go their own way. On my own responsibility I gave orders for lines to be prepared in the rear ready for a retreat. By 1 April I had decided to direct the fighting in such a way that the army could be withdrawn in a more or less orderly manner and without suffering any great casualties, first on both sides of Munster and then behind the Ems Canal and finally to the Teutoburger Forest.

Blumentritt had nine or ten divisions lining the river north of the Ruhr, which in the first week of Operation Plunder had suffered over 30,000 casualties, mostly men taken prisoner.

Surging out of the Wesel-Rees bridgehead, Montgomery's Army Group, including Canadians and Americans, was tasked with striking north-east towards Bremen, Hamburg and the river Elbe, a distance of about 200 miles. On arrival, Kiel was a prime target 100 miles north of Hamburg, and the Danish frontier, 50 miles north of Kiel. Initially Horrocks' 30th Corps and Ritchie's 12th Corps would link up with 6th Airborne and expand and widen the bridgehead for General Barker's 8th Corps.

By the end of March the British Liberation Army was almost in a line abreast. In the far north 49th Polar Bears, with the Canadians, were battling into Holland east of the Grebbe defence line. Two Canadian divisions would form the northern flank of Operation Plunder, then 3rd British, 51st Highland and 43rd Wessex Wyverns, led by the Guards Armoured. Further south the 7th Desert Rats armoured would lead 52nd Lowland, 15th Scottish and 53rd Welsh. The *schwerpunkt* striking force south would be 11th Armoured, occasionally 15th Scottish, with 1st Commando Brigade, who had done so well in Blackcock, with 45th Marine Commando and, of course, 6th Airborne, who had 6th Guards Independent Brigade of Churchill tanks under command.

The whole British Army, apart from the Polar Bears, was involved in Operation Plunder.

After the Rhine crossing the commando units were seasoned water-barrier crossers. In front of 8th Corps were the river Ems, the Dortmund-Ems Canal, river Weser, river Leine, river Aller and finally, the great river Elbe. All of these were ideal defensive barriers.

The 11th Armoured Division soon realised that one of the main weapons of destruction for their new and shiny Comet tanks was the cheap little throw-away, hollow-charge panzerfaust. In bazooka country

Convoy of 'soft' vehicles move into Germany behind the armoured division..

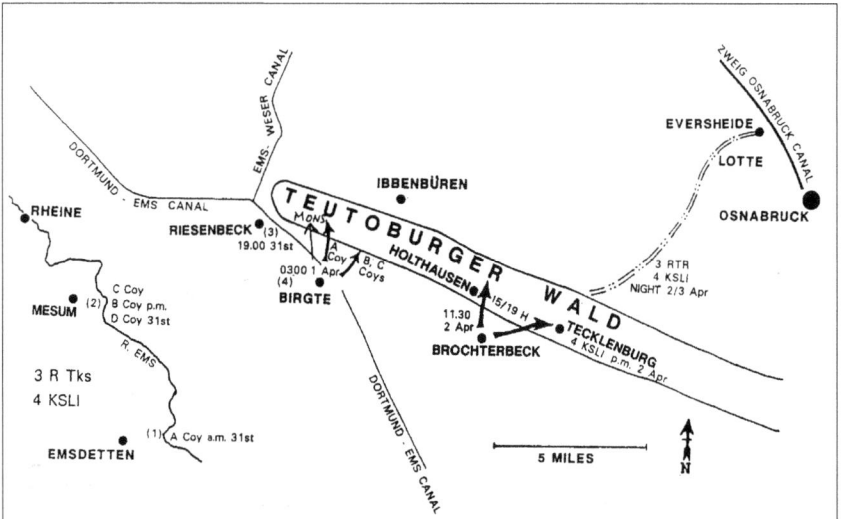

The Teutoburger Wald. Since the time that Caesar's Legions were defeated here, this has proved a superb 30-mile wooded defensive position. 11th Armoured made the first attack on 1 April. The Author's halftrack .50MG was used to shoot down a Focke-Wulf plane!

it was necessary always to carry infantry with the leading squadron of tanks. Unfortunately the shiny new Comets all needed modifications to their driving sprockets and the necessary parts were *eventually* flown out from England a few hours before the tanks were loaded on transporters. By the morning of the 29th, the 'Black Bull' formation was assembled in Wesel and was directed initially on Holtwick, 17 miles north-east, which was captured by 3rd RTR, then Horstmar. All the minor roads were diabolical and continued to be so for the next four weeks; twenty tanks were bogged down in thick mud. On the left, northern flank, the roads were so bad that 7th Armoured were unable to move for two days. The small town of Burgsteinfurt was captured by 4th KSLI on the night of 30 March, the first of many successful night actions by the division. The bridge at Emsdetten, predictably, had been destroyed, but 3rd RTR found a good bridging site at Mesum, 5 miles north.

Between villages, the armour 'brassed' up every wood and bit of cover with Besa machine-gun fire. In Schoppingen, Holtwich and Gessen bazooka parties damaged half a dozen tanks. Fortunately the Comet armour was far tougher than that of a Sherman.

Some 12 miles west of the large town of Osnabruck runs a long strip of denser woodland, two to three miles in width, some 30 miles long,

Ibbenbüren. The Teutoburger Wald ridge was stubbornly defended.

an escarpment with three key villages – Ibbenbüren, Holthausen and Tecklenberg. At its north end it is less than a mile from the canal, and at its south end 10 or 12 miles away.

The journalist and author, Chester Wilmot, described the initial battle:

2 April 1945. Last Saturday evening, just at dusk, tanks of the 3rd Royal Tank Regiment reached the Dortmund-Ems canal, the line on which the Germans were expected to make their next stand. Our tanks had orders to seize a bridge, but as they got within striking range of the canal half a dozen white Verylights shot up into the sky. On this signal German engineers somewhere on the far bank pressed the switches that led to the high explosive charges in half a dozen bridges; and up the bridges went. But our orders were that if the tanks couldn't get a bridge, the infantry of the King's Shropshire Light Infantry (KSLI), who were following close behind, were to get across the canal as best they could and hold a bridgehead on the far side.

They had no storm boats and the canal at high water is 40 yards wide. But when they reached it well after dark they found that the RAF had prepared the way for them; because our bombing had breached the banks farther up, the water-level was only three feet instead of the 20 they'd expected. At one point they found a huge barge was aground, and what's more had swung round so that it lay across the canal forming almost a ready-made footbridge.

Along this barge the KSLI scrambled across. But they hadn't gone far beyond the bank when they ran into some difficult customers – some young Germans who'd just been sent up from an NCOs' training school in Hanover, where they'd been training to become sergeant-majors and warrant officers. Our troops soon found that these young Germans were as tough as sergeant-majors in any army are usually reputed to be.

The Germans moreover had an ideal defensive position. The canal here runs under the lee of a thickly wooded ridge that rises sharply and dominated the whole area. This ridge is part of the famous Teutoburger Wald where the Germans in the past have stopped so many invaders, including Charlemagne. Our infantry soon found they couldn't get on and might even be in danger of being forced back if they couldn't get some heavy weapons across the bridgeless canal that night. The Engineers came to the rescue. Near the barge, where the water was shallow, the canal runs between two high embankments. With an armoured bulldozer, the Engineers pushed these embankments into the bed of the canal, and within a few hours that night they had the tanks across.

Teutoburger Wald – April 2nd, 1945. 2 FF Yeo supporting 3 Mons 11th Armoured Division.

The tanks also towed over some guns, and with this double support the KSLI were able to hold the enemy off all yesterday, while Engineers built a bridge over the canal behind them. This was no easy task, for the bridgehead was under fire and it even drew on itself an attack from the Luftwaffe. The bridge was nearly finished when some German fighter-bombers came over, and they put one bomb less than 20 yards from the bridge where the Engineers were still working; but the bomb stuck in the mud.

With this bridge in action we soon had more troops and tanks across, but the German NCOs on the timbered ridge couldn't be shifted easily. Last night and today infantry of the 11th Armoured Division have been beating through these woods, having almost to drive the Germans from tree to tree; and while the NCOs held out here, some of our reconnaissance tanks [15/19th Hussars] got through along the main road. There they forced a passage up a narrow defile that was defended by a second-rate reinforcement battalion. By getting round this way, they reached the top of the ridge, swung round behind the German NCO battalion, and when I left the front this afternoon its defence was cracking under this sandwich attack.

Meantime, having opened up the road up the escarpment, the 11th Armoured Division was pushing on, rolling up the defence of the bridge by attacks on the flank. At one town they reported they were having some trouble with civilians. Presumably the SS police had been sent into this town also to make the people fight. But now, having broken out on to the plateau behind the Teutoburger Wald, the 11th Armoured is barely 12 miles from Osnabrück. Today's battle yielded them even more than that. Hidden on that wooded ridge they found one of Germany's great secret oil refineries. The loss of this refinery at this stage of the war is probably almost as serious for the Germans as the loss of the Dortmund-Ems Canal.

Chester Wilmot was an optimist. The gallant 3rd Monmouths who had done so well in Normandy were torn apart in the woods, although they acquitted themselves well and in the hand-to-hand fighting Corporal E. T. Chapman won the Victoria Cross. They captured 100 cadets and inflicted many other casualties, but by mid-morning on 4 April they had had forty-one killed and eighty wounded; three consecutive commanding officers had been killed in action. Their C Company had had six consecutive commanding officers killed in action since D-Day, and they had lost 242 killed in action, including twenty-five out of sixty-seven officers; their overall casualties were over a thousand. General Pip Roberts had to withdraw them and the survivors went back to Wesel to guard the bridges.

Desert Rats tanks in Ahaus, probably 1 RTR Shermans. (*Imperial War Museum*)

The north-west sector of the Teutoburger Wald was officially within the 7th Armoured Division sector. The Desert Rats had crossed the Rhine on 27 March, led by the Cherry Pickers, the most experienced of the armoured car reconnaissance regiments in the British Army. They met the 6th Airborne at Hamminkeln. The enemy ahead had been identified as two battalions of 857th Grenadier Regiment and 33rd Panzergrenadier Ersatz Battalion, with infantry, dozens of Czech-made SP assault guns, 88mm anti-tank guns and many panzerfaust teams. The Desert Rats' line of advance was to be towards Bremen via Borken, Stadtlohn, Ahaus and Rheine. They followed the 'Yellow' route to Raesfeld, took Heiden, and encountered bazooka teams in Ramsdorf. Night attacks followed to clear Ahaus, Heek and Nienburg. Major Tom Craig, 1st RTR, wrote: 'We had infantry in APCs and 5th RHA in close support, "artificial moonlight" to light our way, and covered ten miles before dawn. It wasn't plain sailing with opposition from SPs and the inevitable hand-held panzerfaust anti-tank weapons. We sadly lost two tanks.' Major Freddy Pile, 1st RTR, noted:

As the advance started we normally moved down roads or tracks for speed. One troop (a different one each day as it was demanding and dangerous) would be ordered to lead the squadron. Squadron HQ came next and the four

Close support tanks of 15/19th KRH giving infantry support with their 95mm Howitzers, in the Teutoburgerwald battle, April 1945.

A Challenger, slave carrier and Cromwell of 15/19th KRH on the Teutoburger Wald, April 1945.

other (reserve) troops behind. Nearly always we made some contact, perhaps some infantry in personnel carriers with a few motor cycles, possibly two or three tanks or SP guns. With their superior guns they could inflict enormous damage on our Cromwell tanks if these were caught in the open ... We were required to keep going at all costs and enemy resistance had to be overcome without much delay or there would be sharp criticism on the radio from our Brigade HQ, even with heavy losses the instruction from above was always 'Sorry, you must press on regardless'. The phrase 'pocket of resistance' is rather a misnomer: it was usually a small but very well-armed mixed group of tanks, guns and infantry, usually sited on a main approach road, or in a wood or village, difficult for us to circumvent. Because we often ran into them without much warning we had casualties quite out of proportion to the size or importance of many of these actions. As we approached towns or villages we often used incendiary machine-gun ammunition to set fire to the houses in the path of our intended advance. This, of course, did cause much damage.

During the night of 30 March Ahaus was entered – a dangerous wrecked town with many bomb craters, mines and booby traps. On 1 April, Rheine was reached after 20 miles of a real 'swan'. However, the Desert Rats encountered Panzerjagd Kommandos, groups of selected officers and men who stayed behind to act and behave as dangerous 'werewolves'. On the 2nd the Desert Rats crossed the river Ems at Rheine and joined in the attritional battle of the Teutoburger Wald. By then 7th Armoured had advanced 120 miles, almost continuously opposed. 11th Armoured had left behind a battlegroup to hold the bridgehead over the canal. On the 3rd the Desert Rats infantry, 9th DLI and 2nd Devons supported by 5th 'Skins' Dragoon Guards began a grim battle against the fanatical officer cadets defending Ibbenbüren. It failed. On the 4th again it failed. The 1945 official divisional report says: 'The battle continued until dusk with deadly enemy sniping and fanatical resistance from well-concealed positions. Tanks who spotted the enemy in houses burnt them down immediately – but the enemy remained in the blazing ruins firing to the last, before themselves being burnt in the holocaust.' The Desert Rats were ordered to resume their advance.

Some of the SS cadets escaped from Ibbenbüren and specialised in ambushing HQ and B echelon vehicles. They killed Captain Cordy-Simpson, 1st RTR, in the forest of Langeloh. A few days later his friend and commanding officer, Major Tom Craig, wrote: 'We had a lucky stroke of revenge shortly after John was killed. North of Soltau near the village of Jarlingen, I was right behind my leading troop, was astonished to see marching across a T-junction about 300 yards away, a group of about 40 SS cadets marching towards the village towing eight carts laden (we discovered later) with panzerfausts and ammunition.' Major Craig sent a troop of his Cromwells to the left to stop the cadets escaping back to the village, brought up his HQ squadron troop into line. They all opened fire with HE and BESA. 'It was quickly over. There were no SS survivors.'

The 52nd Lowland Scots Division, the Mountain and Flood Division – blooded in Walcheren, matured in Blackcock and Veritable – had moved steadily out of the Wesel bridgehead and headed for the town of Rheine, 25 miles due west of Osnabruck. It is a most unlikely story! Their RAOC unit had a bath unit: a monstrously heavy truck with portable water tanks, furnaces and perforated pipes, which rapidly assembled would give the battle-weary PBI a hot spray/shower/bath. The sergeant in charge managed, by driving down devious side roads – lost perhaps – to enter Rheine without opposition. They found a railway bridge spanning a main street, stopped, set up shop, got the water hot, even boiling, and the fires were burning nicely, ready to provide hot showers. Rheine, of course, had not been captured. An astonished platoon of Lowlanders advanced ready for street fighting and were invited to have a hot shower before finishing their attack and capturing the town! A substantial battle followed, helped by 4th Armoured Brigade and the town was taken by midnight on 2 April.

On 1 April the Lowland recce found the bridges blown along the river Ems, and identified their German opponents as the 60th Panzergrenadiers, the 1st Battalion Grossdeutschland Regiment, plus NCO candidates from the Hanover training establishment. Major General Hakewell-Smith now had to force a crossing of the Dortmund-Ems Canal, where all three bridges were down. The canal had high steep banks. It was an area of poorish farmland, ditches, wire fences, winding streams, hedges, farm tracks, thick copses and spinneys, small farm houses, crofts and, above all, a sodden countryside difficult for tanks and carriers. On 4 April 7th Cameronians forced a canal crossing and 5th Highland Light Infantry widened and extended it. Their objective was Dreirwalde, three miles east. Hungarian SS troops lost 400 men in its defence, mainly to the Glasgow Highlanders.

The next objective was a night attack by the RSF and 6th Cameronians on Hopstein. George Blake, the divisional historian, who was there, described:

> ... the eerie journey of exquisitely cautious movement, of likely surprises from the many streams, crawling in the darkness. At dawn a good old-fashioned infantry charge with fixed bayonets, 'the blood of the Covenanters was up' as they killed or wounded all the defenders. Eight miles south of Hopstein was the Teutoburger Wald ... the finest defensive position, two long heavily wooded ridges – the most notable feature of the expanse of the flat lands of North Germany. Local quarries provide a fine yellow sandstone. Solid houses built of this material please the eye, with relief from the monotonous red brick of a thousand red brick hamlets. Between is the rather dull Westphalian Plain. Ibbenbüren lies cosily between the two spurs of the main ridge, a comic, contorted little place in the Walt Disney tradition ... looking amiably towards the fields that slope down from the western spur.

Brigadier McLaren's 155th Brigade was left to deal with the town, which has a tortuous defile a mile to the west similar to a Bournemouth or Sandown chine. 7/9th Royal Scots and 6th Highland Infantry attacked on 4 April from a cement factory and quarry. By nightfall the wooded ridge was secured. One 6th HLI platoon commander said: 'This is one hell of a place. We have been driven off here [like the 3rd Monmouths two days earlier] once already. We've been counter-attacked all night. Good luck and goodbye.' Indeed 7/9th Royal Scots, 6th HLI and finally

'Flying Dustbin' Spigot mortar bomb fired by an AVRE to demolish a road block in Bocholt, 29 March 1945. (*Brian Haward*)

4th KOSB all took casualties before the town was cleared for 53rd Welsh Division to pass through and mop up.

The 53rd Welsh Division had had a very bad week in Blockbuster trying to capture Weeze in their local operations Leek and Daffodil. They crossed the Rhine on 26 March opposite Vynen, then crossed the river Ijssel just west of Ringenberg, Dingden and had a battle to capture Bocholt, an important road centre with many narrow streets, road blocks and minefields. They were helped by 4th Black Rat Armoured Brigade and by the AVRE bombards on the Churchill tanks of 82nd Assault Squadron RE. The 4th RWF mounted in Kangaroos fought four successful actions on the Hohe Heide overlooking the town. Eventually 1st HLI cleared the final opposition in the northern factory area during the night of 30/31 March. It was a curious route for the Welshmen along the German-Dutch border; in Winterswijk and Bredevoort in Holland there was a great patriotic welcome with flags, bunting, cheers, smiles and happiness. In Vreden, Rhede, Stadtlohn and Alstatte, the inhabitants were sullen and dispirited in their smashed German villages. Next came Gronau with the first of many large hospitals containing 700 German wounded soldiers. Then the Rheine aerodrome was cleared by 2nd

Monmouths. By 5 April the division was established on the line Gronau
– Ochtrup – Saltzbergen and their new role was to put two brigades
to mop up the Ibbenburen area. In the four weeks up to 7 March the
53rd Welsh had 3,700 casualties, of whom 20 per cent had 'battle stress'.
Supported by 7th RTR Crocodiles, 71st and 158th Brigades mopped up
the town and area. The RAF had bombed and medium and heavy artillery
bombarded the cadet companies and still they hung on defiantly.

Desmond Milligan's Oxfordshire and Bucks carrier group helped clear
Ibbenbüren town for the final time:

> We made a house-to-house search through this now battered little town,
> constantly dodging the shells lobbed into the area by artillery supporting
> the cadets. About eight enormous very tall trees had blocked the road and
> stopped the tanks the previous night. An armoured bulldozer came up to move
> them and our carriers were given the job of protection. Two German officers
> carrying a white flag appeared. Lt Hawley led two of us to take possession of
> a hospital full of Allied wounded [11th Armoured, 7th Armoured and 52nd
> Lowland]. A ragged cheer went up – the wounded lay and sat everywhere so
> heavily bandaged you could hardly see them.

On 7 April Lt Colonel Burden, CO the East Lancs, wrote: 'Only 21 PW
taken. The Nazi cadets seemed to prefer death to capture, exposing
themselves to the full blast of MGs or flame jets from the crocodiles, as
they chanted Nazi slogans to the last, obstinately determined to die for
Adolf Hitler. The party propaganda worked on impressionable young
minds.'

The battle of Ibbenbüren and the Teutoburger Wald had taken a week
of incessant close-quarter fighting. Four of Montgomery's finest divisions
had taken it in turns and had suffered in the process. The armour of
course were tasked to get *through* the Teutoburger Wald and to continue
to smash their way to Bremen and Hamburg.

CHAPTER 16

T-Force: 'Thieving Magpies'

The British Army is usually composed of regular regiments with soldiers making a military career. In the First and Second World Wars the vast majority of formations were intended to be active only for the duration of the war. But there were also a number of 'irregulars'. The classic was 'Popski's Private Army', but also there were Phantom, the secret radio sections, and 30th Assault Regiment, which specialised in retrieving secret naval information and technology from captured enemy ports and harbours – and then there was British Target Force.

The western Allied leaders knew that the Third Reich was developing many forms of technical warfare – particularly in rocketry, jet aircraft, deadly new gases, highspeed submarines, infra-red gun sights, radar systems and possibly nuclear weapons, and other, totally new deadly concepts. In July 1944, General Eisenhower issued a secret order from SHAEF to raise a 'Target-Force'. Lt General W. B. Smith, on 27 July, ordered senior commanders to raise a secret force to secure vital intelligence and scientific targets, not only the weaponry but if possible to capture their designers too. So T-Forces, American, British and Canadian, came into being, that would operate *independently* from the regular armed forces. Serious planning started in August. In the National Archives at PRO Kew, document WO/219/818 has the original brief: 'T-Force = Target Force, to guard and secure documents, persons, equipment with combat and intelligence personnel, after capture of large towns, ports in liberated and enemy territory.'

Commander Ian Fleming RN, later to be famous as the author of the James Bond books, had developed an Intelligence Assault Unit (IAU) that was described in ADM 223/500 as 'a force of armed and expert authorised looters', and was renamed 30th Assault Unit. It served in North Africa, Sicily and Italy with impressive results including uncovering German Enigma code machines. 30th AU also did yeoman work in France and in Belgium, discovering 12,000 important intelligence documents.

Germany would be the target area designated for T-Force activities, and Brigadier G. H. C. Pennycock, head of the Chemical Warfare Group whose staff was seriously underemployed, had the resources to be the nucleus for the new T-Force. His second in command was Lt Colonel R. Bloomfield, and FO 1031/49 contains their secret tasks. The first troops allocated were six companies of Pioneers, whose main task in operations Veritable and Plunder was to create smokescreens. They had forty-nine trucks and trailers. The first infantry formation to join was 5th Kings Regiment, a D-Day 'Beach Group' unit under Lt Colonel B. D. Wreford-Brown. It was deployed to Antwerp in February 1945 as a mobile reserve, and in March took part in the Dunkirk 'blocking' operation. The next unit to join was 1st Buckinghamshire Battalion of the Ox and Bucks Light Infantry, and then came No. 19 Bomb Disposal Section RE, plus many interpreters speaking English, Dutch, German and some Russian. T-Force then mustered itself in Venlo, a large Dutch town on the river Maas. Finally in March the 400-strong 30th AU joined and the combined unit moved to a huge steelworks in Krefeld, east a few miles into Germany. Scientists joined this curious group. Sean Longden, in his evocative book *T-Force* noted the composition of the 'authorised looters': 'A strange mixture of D-Day veterans, teen-age "virgin" soldiers, ex-Royal Marines, former gunners, soldiers returned straight from hospital and shell-shock cases, but they were to embark on a journey that would have enormous repercussions for the postwar future of the whole western world.'

An immense amount of work had been done by a recently formed intelligence section of the Buckinghamshires. Map boards pinpointed which targets would be acquired and checked once across the Rhine. There were three key categories: factories and plants manufacturing military equipment; government and military HQ buildings and staff; military R & D buildings. The capture of key personnel was paramount and of course the preservation of the enemy facilities, and the latter possibility of sabotage or booby traps.

At long last T-Force, plus 150 target investigators, followed the armoured columns swarming over the Rhine and in two days, led by the

Officers of 5th Battalion the Kings Regiment (T-Force) May 1945. Lt Colonel Guy Wreford-Brown, the CO, is fifth from right on the front row.

Kings troops, were on their way into the Third Reich. Splintered into small groups, they were controlled by GSO2 at Army HQ. Targets were made from town plans and maps and aerial photographs. In Osnabrück, the Kupfer und Drahtwerke factory yielded new metallurgical processes for submarine parts. In Hengelo, Holland, they found new types of anti-aircraft predictors. In Rheine they found a vital main telephone exchange and the Takke Fabrichen plant, making U-boat gearboxes, and a huge stockpile of ball-bearings. In Cologne in the US zone, a factory was captured, which produced equipment for guided rocket systems and jet- and rocket-propelled aircraft.

Field Marshal Montgomery's Chief of Staff, Major General de Guingand, issued an official pass or *laisser passer* to ensure that T-Force personnel were not only allowed free, unimpeded access to these targets, but also were permitted to secure and defend the targets. In the event, one of the key problems was the tens of thousands of displaced persons of all races who, once freed from captivity, wrecked and looted anything that took their fancy. Unfortunately, in the American sector, T-Force were often treated with deep suspicion, as an illegal private army seeking profitable 'loot'.

The military area of Wehrkreis in Hannover, and the huge Hanomag truck factory producing radar equipment using sintered iron instead of copper, was inspected. The Maschinen Fabrik disgorged infra-red technology and Professor Hase, a distinguished physicist. Doctor Max Kramer produced secret X4 and X7 rocketry of great interest. The Continental Gummiwerke specialised in synthetic rubber. Often the T-Force investigators were the first

Above: A jeep of B troop, 30 AU, Royal Marines at Minden, Germany, May 1945.

Left: T-Force Royal Navy investigator John Bradley (left) with Commander Dunstan Curtis of 30 AU. Bradley was a metallurgist.

T-Force evacuating equipment from a factory in Köln (Cologne), March 1945.

to enter towns such as Delmenhorst and Bomlitz before the regulars turned up. In Bomlitz, a Focke-Wulf aircraft factory and a helicopter factory were captured. So it went on: diesel submarine engines in Twistringen; radar assemblies in Nienburg; a natural gas field in Bentheim; marine mines to be fired by U-boats in Starkshorn; a Krupps test area in Meppen; a Blohm and Voss aircraft factory in Wensendorf; a huge underground factory with two miles of Focke-Wulf aircraft parts in Celle near Belsen.

Major Brian Urquhart, the discredited Intelligence Officer in Market Garden, had joined T-Force. He was responsible in late April for capturing professors Groth, Suhr and Faltings, distinguished scientists working on the isolation of isotopes of uranium for nuclear purposes. They and their crated-up laboratory were flown back to the UK under guard. And almost by mistake Professor Otto Hahn, the distinguished physicist at the Technische Hochschule in Hannover, and the chief scientist for atomic research in Germany, joined the others in a benevolent captivity. Rather unpleasant chemical weapons were discovered at Hunstadt, Frankenberg, Rehden and Espelkamp. These unknown gases and filled shells were sent, very carefully, to Porton Down research facility. Their staff flew out to Munsterlager, the major gas munitions centre in Germany, for chemical warfare. At Bassum near Bremen, naval fire-control apparatus and torpedo predictors were discovered dispersed in barns, farms and houses.

'The Thieving Magpies', as they were rudely but accurately described, will turn up again in this book.

CHAPTER 17

Bergen-Belsen
Concentration Camp

Major General Pip Roberts, GOC 11th Armoured, after leaving the
terrible Ibbenbüren battlefields to be cleared by the follow-up infantry
divisions, wrote:

> Our next obstacle was the Osnabrück canal and the bridge over it at
> Eversheide, north of Osnabrück. The clearing of Osnabrück [a significant
> town] was the task of 6th Airborne Division who were keeping abreast of us,
> almost impudently as it seemed to us, for apart from their own recce regiment
> they had only a Churchill tank regiment from 6th Guards Tank Brigade,
> excellent machines but far outpaced by our [brilliant, shiny, new] Comets. Yet
> every day they inched up on our right and even now as we were moving to cut
> off Osnabrück from the north, their leading troops were within a few miles of
> its southern perimeter. Obviously we must do another night 'march' as we had
> done at Amiens with 3rd Royal Tanks in the lead. But the way was not easy.
> The Germans resisted by night as they had by day and several defended road
> blocks had to be cleared before the canal could be reached.

Actually 6th Airborne had had a two-day head start in Operations
Plunder and Varsity and they had not received a bloody nose by the
Hannover cadets in the Teutoburger Wald. The two bridges at Eversheide
were seized intact and the race was on!

Wynford Vaughan-Thomas reported on the right flank advance:

> 7 April 1945. Our 11th Armoured Division, with the 6th Airborne on its
> southern flank, has raced down the main roads from Osnabrück towards the
> river Weser. As they went they threw out little parties to form roadblocks on

Crusader Comet
tank at Neustadt
on the way to
Bergen-Belsen. (*Tank
Museum*)

the side roads to the north and south of them, just in case any stray Germans should come in on the flank. And the airborne boys travel in the most varied collection of transport I've yet seen, in fact they can't call themselves airborne any longer, they're either lorry-borne or tank-borne, motor-bike-borne, even cart-borne infantry. The Germans have little before us now to stop our racing columns. They rushed some units up, but their trouble is that they're at least 24 hours behind in their information. One of our columns bumped into a regiment that had come down from Denmark. Apparently these Germans were on their way to defend Osnabrück, oblivious of the fact that it had already fallen days before. We caught one battalion of it just as it was moving into billets in a small village. Well, those Germans are now billeted in our prison cages.

But speed brings with it its own problems. Apart from that ever-present nightmare of supply, how to get the petrol and the food up to our leading tanks, there are one or two other headaches for our Army. There's the problem of the prisoners, for example. What on earth are you to do with these thousands upon thousands of shabbily uniformed, middle-aged, and undersized warriors that are all the High Command has got left with which to try and stop our armies. For now we really are coming to the bottom of the bag. The German manpower is exhausted. But here are the survivors of warfare, cluttering up the roads and offering to give themselves up to the first person who will have

221

them. It's difficult for an armoured division to begin to cope with them. Our tanks have got to get on and our lorries are wanted to bring up supplies. The only thing to do with the prisoners is to disarm them and send them off to the rear. And with an advance as fast as ours the rear may be 60 miles away. These are just the ordinary spectacles – miles behind the line, long columns of prisoners, sometimes led by their officers, quietly tramping along in the middle of our back areas. And with nobody paying the slightest attention to them. And with none of them making the slightest effort to escape.

Most divisions had increased their mobility and manpower by making alternative use of their surplus artillery formations. 11th Armoured called theirs 'Todforce' as the CO of their anti-tank regiment, Lt Colonel A. F. Tod, redeployed his unit as two squadrons of 17-pdr 'tanks' and two companies of mobile infantry, whose towed anti-tank guns were put into store.

The next difficult river barrier was the Weser. On the way Stolzenau and Schlusselburg were captured, but both bridges, predictably, were blown. The bridging operations were rudely interrupted by a rare Luftwaffe air attack by Junkers 88s and Messerschmitt 109s. They were seen off by Tempests and Typhoons, but the Training (Ersatz) Battalion of 12th SS Panzer Division and 100th Pioneer Brigade put up strong resistance. A sergeant of the 12th SS described the battle for Stolzenau:

> I was a sergeant instructor in the battalion. My own service had been with the Leibstandarte in Russia. Usual decorations, Iron Cross, Infantry Assault, Close Combat, wound badge in silver. I lined my platoon along the railway embankment. This stands about 20 feet above the ground between the river Weser and the village of Leese. The first British probes were easily repulsed – our field of fire was excellent. We could see every movement that the Tommies made. I had my mortars behind the railway embankment with the Reserve sections and three, tripod-mounted MG 42s firing on fixed lines. The battalion had ammunition enough. We had been ordered to hold to the last. We did not know it then but there was a V-2 assembly plant about two kilometres behind the railway line, just north of Leese, and it was our task to hold until the civilian scientists had got away.

1st Commando Brigade arrived and a triple attack was made: the commandos against Leese, and battle groups of 15/19th Hussars and

1st Cheshires over the bridge at Petershagen (lent to 11th Armoured by the 6th Airborne) and by 23rd Hussars bombardment from the west bank. An enormous quantity of good German hock was discovered in Stolzenau, to everyone's delight! Once over the Weser, Todforce was tasked with guarding explosives factories, some V-2 sites, and large dumps of gas shells in the many woods. 11th Armoured's battle groups pushed hard to the next line of defence held by 12th SS, the river Aller.

The town of Winsen was well defended but over 12–13 April, it and the villages of Walle, Bonstorf, Bergen and Hetendorf were taken with several hundred prisoners captured and a dozen anti-tank guns destroyed. The 11th Armoured's battle groups of tanks and infantry were working well.

During the night of 11 April, 4th KSLI crossed the river Aller to help widen the commando bridgehead. The author was FOO with 4th KSLI and crossed the wide river in one of several black rubber boats. He called up defensive fire that night as young German marines launched attacks in thick woodlands. His Sherman tank was the first to be ferried across on sapper rafts the next morning. His OP was attacked on the morning of the 12th by a Tiger tank and supporting cast of marines, which was distinctly unpleasant. 13 April was 'Black Friday'. 3rd RTR had many encounters with Tiger tanks, which they rarely won. Many Comets were bogged down in streams and marshes – sitting targets. All four of the 13th RHA FOOs were in trouble on the way to Winsen, which was defended by an anti-tank officers' training unit equipped with 88mm, 75mm SP guns and many panzerfaust teams. All day long the slow, tedious and painful battle to clear woods continued. Typhoons in 'Limejuice' targets helped, but 11th Armoured casualties mounted.

Knowledge about the appalling concentration camp of Bergen-Belsen only slowly came to light. On 12 April at Buchholz, preliminary negotiations started when the commander of the German garrison troops approached 159th Infantry Brigade HQ. The Germans were blindfolded and sent back to 8th Corps HQ. In exchange for an exclusion zone for medical reasons, the German garrison would give 11th Armoured free passage, including access over the bridge on the river Aller at Winsen. Negotiations went on back and forth until a local agreement was concluded, fixing a limited typhus zone around the camp itself. The enemy then blew the Winsen bridges. Alerted to the danger of typhus, every member of the division was doused by the RAMC with white

SS camp guards amid hundreds of
dead prisoners – Bergen-Belsen.

DDT powder from head to food; it was sprayed inside shirts, trousers
and boots. Very early on the morning of the 15th, a Commando jeep
sped through the camp, then some Inns of Court armoured cars, and the
23rd Hussar Comets and 3rd RTR with 29th Armoured Brigade HQ
under Brigadier Roscoe Harvey. The author, as an FOO with 3rd RTR,
was in a Sherman tank and as chance would have it, his part of the long
armoured column halted for about two hours in the centre of the camp.
The minor metalled road divided the men's camp from the women's.

No one had come across such appalling evil before. The smell and
stench of decomposing bodies hit one. Behind the high wire fences the
grounds were covered in tattered grey rags and bundles of the dead or
dying. It was difficult to detect any movement. Immediately on the
right, most of the forty-four German/Hungarian staff were lined up,
quite casually looking at the overwhelming forces in their midst. The
infamous scar-faced Colonel Josef Kramer, a professional SS gaoler from
Auschwitz, was the male camp commander. Irma Grese, a tough, tall
blonde SS woman aged twenty, his assistant and reputedly his lover, was
the manager of the women's camp. Originally the joint camps were meant
to hold 10,000 inmates. On 15 April 1945, they held 60,000 (35,000
men and 25,000 women). What this author and the armoured column

saw was 13,000 unburied corpses. One looked at Kramer and Grese and they looked back with mild curiosity, certainly not fear or sorrow. They knew the British would keep their word. They had 24 hours' grace to scarper. About eight tanks behind the author's Sherman was his CO, Lt Colonel Robert Daniel, a professional soldier who had commanded 1st RHA in North Africa, earning him the DSO. He came from a distinguished Jewish family. After a short time there was the sound of shots. The colonel had climbed down from his Sherman, walked into the camp and shot four of the camp guards. He came back and said to the brigadier: 'It's not much, but it is a token of my disgust and horror.' Bergen-Belsen was not an official extermination camp like Auchswitz. From the beginning of 1945, tens of thousands of Jews had been brought from the slave labour camps of Silesia and Eastern Europe. The Germans did not want them to fall into the hands of the Russians, but at Belsen there was no work for them to do, unlike at the vast slave labour camp 100 miles further south, at Nordhausen. Once at Belsen, prisoners who had already survived three, four, or even five years of the rigours of camp life elsewhere, were left, with almost no food or medical help, to starve and to rot.

Bergen-Belsen was originally a detention camp holding prominent Jews who, in theory, were to be exchanged for German nationals abroad. It was also classified as a *krankenlager*, a reception camp for sick prisoners. From November 1944 conditions deteriorated rapidly. There was evidence that Kramer was selling food rations meant for the prisoners. An outbreak of spotted fever caused Kramer to ask for closure of the camp. The newsreel cameras showed the 'Beast of Belsen': stocky and well-fed among the dying and the dead, his cheek scarred beneath the dark stubble.

Every camp in Germany had a *totenbuch* (a death book) giving records of inmates and their deaths. In Belsen, 7,000 died in February, 18,000 in March and 9,000 in the first two weeks of April. There were no facilities for burial; most of the corpses were piled up randomly around the camp. There was no water and no sanitation. Dysentery and typhus were rampant. And despite the very best efforts of the British Army Medical Corps, which included Lt Colonel J. A. D. Johnson, captains Derek Sington, R. Barber and Dr Rosensaft, and Brigadier Llewelyn Glyn Hughes, in the next eight weeks 13,944 prisoners died.

Poor little Anne Frank, the intelligent fifteen-year-old Dutch girl, and her sister Margot, aged eighteen, both died in Belsen in March a few weeks before the British arrived.

Anne Frank.

Appendix O to Chapter VII of 2nd Army History deals extensively with the Belsen camp, and the following extracts are taken from it:

Disease of all kinds was rife and in a vast number of cases it was difficult to tell which condition predominated – whether it was typhus, starvation, tubercle or a combination of all three, which was responsible for the shattered wrecks of human beings who formed the majority of the inmates ...

It was impossible to gauge the daily rate of new cases (of illness) as so many, if too weak to call for help, received neither food nor treatment ... Conditions in the huts were indescribable. Rooms and passages were crowded with both living and dead; only by entering the huts could the real horrors be appreciated ... the appalling sanitary conditions in which excreta from those too weak to move or help themselves fouled the rooms or trickled through from upper bunks to those below ... Latrines were practically non-existent and what there were consisted simply of a bare pole over a deep trench without any screening ...

There had been no water for about a week owing to damage by shell fire to the electrical pumping equipment on which the system depended. Food was of poor quality and the number of meals varied from one to three per day. During the last days (of German control) the entire rations issued were one bowl of soup per person on each of four days and once only a loaf of black bread, divided between several people ... There was no shortage of food stocks from which supplies could have been drawn ... Red Cross boxes were found containing tinned milk, soups, meats, etc, which had been sent by Jewish societies for Jews; these had been stolen by the guards and ... nothing had been distributed.

As a first step the British Army sealed the camp using men from one of the batteries of 63rd Anti-Tank Regiment. A Field Hygiene Section was sent in by 11th Armoured Division and twenty-seven water carts supplied fresh water in place of the camp's infected supplies. So bad had been the mismanagement of the prisoners that during the first few days of British control, 548 of the inmates died in a single day. But within a short time the British Army had the situation in hand. The typhus-ridden were isolated; diets were provided to ensure that the camp inmates did not die from over-eating, the dead were buried and the area was disinfected.

Josef Kramer's staff of forty-four included Dr Klein, who headed the camp administration and also invented tortures, one being injecting creosote and petrol into the prisoners' veins. The staff included sixteen SS men, sixteen female SS guards, and twelve KAPOs including six Poles. Camp 1 (men) was 2 km distant from Camp 2 (women).

The conditions that the British soldiers encountered as they entered Belsen on 15 April were carefully and meticulously documented by the British medical personnel who arrived in the camp in late April. An investigation carried out by Lt Colonel F. M. Lipscombe of the Royal Army Medical Corps reported that camp inmates had been subsisting since January on a daily diet of 300 grams of rye bread, watery soup, and a root vegetable called mangold wurzel, a cousin of the beet and normally used as cattle feed. As in all concentration camps, 'what each individual actually received depended mainly on his ability to obtain it' – that is the weak and feeble went without. 'The great majority of the internees had received no food or water for some five days before the camp was uncovered.' The inmates suffered from scabies, dysentery, sepsis of sores and wounds, typhus, tuberculosis, and the debilitating effects of prolonged malnourishment. The psychiatric scars were also visible. According to Lipscombe, 'the loss of moral standards and sense of responsibility for the welfare of others was widespread.' The normal human 'fear of death and cruelty was blunted by repeated exposure – this especially noticeable in children'.

The concentration camp at Bergen-Belsen, liberated by the British on 15 April 1945. The photograph shows a soup kitchen operated by British soldiers.

An exhausted and emaciated prisoner near death in liberated Belsen. (*Imperial War Museum*)

SS Colonel Josef Kramer, styled the 'Beast of Belsen' in the British press, shown in a mug shot while awaiting trial. He was hanged in December 1945. (*Imperial War Museum*)

7 RTR Crocodile, belching flame, clearing fouled ground, Belsen Concentration Camp, 16 April 1945.

There are many true and totally appalling accounts of the horrors of that evil place. This is Richard Dimbleby's account for the BBC:

19 April 1945. I picked my way over corpse after corpse in the gloom, until I heard one voice raised above the gentle undulating moaning. I found a girl, she was a living skeleton, impossible to gauge her age for she had practically no hair left, and her face was only a yellow parchment sheet with two holes in it for eyes. She was stretching out her stick of an arm and gasping something, it was 'English, English, medicine, medicine', and she was trying to cry but she hadn't enough strength. And beyond her down the passage and in the hut there were the convulsive movements of dying people too weak to raise themselves from the floor.

In the shade of some trees lay a great collection of bodies. I walked about them trying to count, there were perhaps 150 of them flung down on each other, all naked, all so thin that their yellow skin glistened like stretched rubber on their bones. Some of the poor starved creatures whose bodies were there looked so utterly unreal and inhuman that I could have imagined that they had never lived at all. They were like polished skeletons, the skeletons that medical students like to play practical jokes with.

At one end of the pile a cluster of men and women were gathered round a fire; they were using rags and old shoes taken from the bodies to keep it alight, and they were heating soup over it. And close by was the enclosure where 500 children between the ages of five and 12 had been kept. They were not so hungry as the rest, for the women had sacrificed themselves to keep them alive. Babies were born at Belsen, some of them shrunken, wizened little things that could not live, because their mothers could not feed them.

One woman, distraught to the point of madness, flung herself at a British soldier who was on guard at the camp on the night that it was reached by the 11th Armoured Division; she begged him to give her some milk for the tiny baby she held in her arms. She laid the mite on the ground and threw herself at the sentry's feet and kissed his boots. And when, in his distress, he asked her to get up, she put the baby in his arms and ran off crying that she would find milk for it because there was no milk in her breast. And when the soldier opened the bundle of rags to look at the child, he found that it had been dead for days.

There had been no privacy of any kind. Women stood naked at the side of the track, washing in cupfuls of water taken from British Army water trucks. Others squatted while they searched themselves for lice, and examined each other's hair. Sufferers from dysentery leaned against the huts, straining

Burning the last hut at Belsen, 21 May 1945.

THIS IS THE SITE OF
THE INFAMOUS BELSEN CONCENTRATION CAMP
· LIBERATED BY THE BRITISH ON 15 APRIL 1945 ·
10000 UNBURIED DEAD WERE FOUND HERE,
ANOTHER 13000 HAVE SINCE DIED,
ALL OF THEM VICTIMS OF THE
GERMAN NEW ORDER IN EUROPE,
AND AN EXAMPLE OF NAZI KULTUR.

A sign erected at Belsen.

helplessly, and all around and about them was this awful drifting tide of exhausted people, neither caring nor watching. Just a few held out their withered hands to us as we passed by, and blessed the doctor whom they knew had become the camp commander in place of the brutal Kramer.

I have never seen British soldiers so moved to cold fury as the men who opened the Belsen camp this week, and those of the police and the RAMC who are now on duty there, trying to save the prisoners who are not too far gone in starvation.

It is an evil story and soon the whole world knew about Belsen, certainly the whole of the English-speaking world.

There are various endings to the story. In the 24 hours after the truce ended, the camp guards and also Kramer and Grese were captured. They were tried appropriately and justly at a war crimes tribunal and were hanged.

Later in this book there is a chapter on war crimes tribunals. The author served as the junior member of those in Hamburg and Oldenburg; the first aged twenty-one and the second aged twenty-two. He was one of the official British Army of the Rhine observers when Mr Alfred Pierrepoint hanged thirteen camp guards at Hameln on Friday 13 December – before lunch.

On 21 May 1945 flame-throwers burned down the last of the hundreds of wooden barracks at Belsen.

CHAPTER 18

Operation Forrard On to Bremen

Lt General Brian Horrocks' 30th Corps was given the task of heading for Bremen, on the northern flank of Montgomery's great surge towards the Baltic; it consisted of 43rd Wessex Wyverns, 3rd British and 51st Highland divisions. Behind the infantry in the early stage would be Guards Armoured and, initially leading the way, 8th Red Fox's Mask Armoured Brigade.

Once across the Rhine at Rees and Haldern, the 51st Highland would take the right, southern flank, the Wyverns in the centre and 3rd Canadians on the left, northern flank. The 3rd British Division under Major General 'Bolo' Whistler would follow up behind the Guards on a route following the Dutch-German border via Aaltern Groenlo, Haaksbergen, Enschede to Oldenzaal.

Captain Guy Radcliffe, adjutant of 2nd KSLI, wrote:

The area on the east bank of the Rhine had been utterly pulverised by the bombing and the artillery fire. Dead cattle surrounded every derelict farm, the debris of war lined the roads with the torn-off boughs of the trees. The Germans moved about in a dazed state ... we had very little trouble with them ... of repentance there seemed to be none, only a grovelling self-pity. On 1st April the Bn arrived at Lichtenvoorde in Holland. We were overjoyed to be back in friendly country. One could feel the 'freedom' in the air, every house had put out the flags which had long been concealed. We reached Enschede the next day and experienced for the first and last time in the campaign what a large town did when it was liberated.

43rd Division Wessex Wyverns on the march. (*Stan Procter*)

On 30 March Operation Forrard On started with the 43rd Wessex Wyverns and 8th Red Fox's Mask Armoured Brigade on the 250-mile advance towards Bremen. The Staffordshire Yeomanry DD Sherman tanks 'swanned' across the Rhine, almost three-quarters of a mile, and pushed through the 51st Highland Division towards Speld Dorp and Bienen against the old enemy, 15th Panzergrenadier Division. The Sherwood Rangers Yeomanry crossed the river by raft on the new RE bridge to help the Black Watch and Seaforth Highlanders capture Ijsselburg.

Captain John Stirling, 4/7th Royal Dragoon Guards, wrote: 'It took the best part of a week and the names of Rees, Bienen, Millingen, Megchelen, Ijsselburg all represented a stiff tussle before they became ours. We were still battering our heads against the paratroops and they fought with a grim hopeless bravery that no man could fail to admire. To defeat them was to kill the majority and then a few might surrender. In Megchelen, they even continued to fight from houses that we had set on fire with Crocodiles. Fanatical and misguided yes, but brave and well disciplined, also yes.'

Lingen, a sizeable garrison town about 15 miles east of the Dortmund-Ems Canal, was the next task. It was defended by 111th Battalion

German prisoners *en route* to the POW cages.

Grossdeutscher Brandenburger Training Regiment backed up by 700 paratroops. Guards Armoured had progressed well from Groenlo, Borculo, Enschede and Hengelo (all in Holland) and back into Germany to Gildehaus and Bentheim, and had made a daring 12-mile assault through the night to try to seize the bridges over the river Ems at Lingen. Despite capturing three bridges in Nordhorn they were foiled at Lingen. However, a mile or so north, Captain I. O. Liddell won the Victoria Cross by climbing a roadblock and cutting the wires connecting the demolition charges at Altenlingen.

3rd British Division had been trailing Guards Armoured and were tasked with the capture of Lingen, so 2nd KSLI arrived and in total silence at 0230hrs on 4 April crossed both river and canal with rafts and boats and formed a bridgehead. Captain Marcus Cunliffe wrote in the history of 2nd Royal Warwicks:

The attack [from the north side at 1015hrs] was pressed with admirable speed and success. Field Marshal Montgomery's men fought like veterans. Many of them actually were reinforcements, youngsters who had just recently joined. At mid-day the 1st Norfolks came through us [against several hundred paratroops] and fought through the streets, house by house and block by block

235

2nd Gordon Highlanders
cross a temporary
bridge north of Celle,
12 April 1945.

until evening. Then in turn the Lincolns who cleared some houses in the dark.
Street fighting is a laborious business and nerve-racking. Lingen was a large
town.

It was defended by paratroops, SS, Wehrmacht and the Volksturm (Home
Guard). Help soon arrived; the Buffs sent a squadron of Crocodiles who
helped the Lincolns clear many streets. A squadron of Staffs Yeomanry
Shermans and a number of assault gun SPs led several counter-attacks,
and twenty Luftwaffe Messerschmitts suddenly swooped down out of
the blue. Fighting went on until 6 April when Lingen was finally cleared.
Charles Graves, 2nd Royal Ulster Rifles, notes: 'This final phase of
Lingen produced bitter, lethal fighting. The fanatical Germans [of 7th
Parachute Division] asked no quarter and received none. Hand-to-hand
fighting took place and the bodies of the Germans lying within a few
yards of our wounded and dead testified to the fierceness and severity of
the fighting.'

The Wessex Wyverns were given the task of keeping open 'Club Route'
north from Anholt into Holland to the Twente Canal at Lochem, east to
Hengelo, and back into Germany at Nordhorn and Lingen. The tough,
irascible Major General Ivo Thomas created his own Armoured Thrust

Royal Norfolks take prisoners in Lingen. Left to right: Pte Wilby, Pte Phillipson, L/Cpl Gould.

Group, led by 8th Armoured Brigade, followed by mobile groups from 129th, 214th and 130th Infantry Brigades. Lt Sydney Jary, 4th Somerset Light Infantry, an intrepid young infantry leader who wrote *18 Platoon*, recalled: 'It was Bank Holiday spirit until the leading Sherwood Rangers tank encountered the first enemy roadblock covered by 88mm guns or SPs or two or three Spandau teams. We had three nasty little battles at Sinderen, Lochem and Osterlinden. In Sinderen the leading Sherman was brewed up by a Mk IV SP hidden in a haystack. Its passengers were killed by Brens, but 20 Germans charged us with bayonets shouting. Brave lads, they didn't stand a chance. My only order was "cease fire".'

Captain Hancock, an FOO with 4th SLI noted:

As the leading tanks were entering Ruurlo, 6 miles short of Lochem, they were fired upon by Panzerfaust and Spandaus and two of them were hit and burst into flames. The only building nearby bore a large red cross. It was immediately treated to concentrated fire from BESAS (tank MG) and 17-pdrs. Out of it came one German officer and 56 soldiers. It was the Flak HQ of the area containing much transport and six 20mm guns. We reached Lochem by 1000hrs on 1 April. The road between was littered with shot-up trucks, dead horses and waggons, assorted staff cars, some with their bullet-riddled occupants still sitting as they were when the RAF [Typhoons] caught them fleeing to the rear.

The Guards captured Enschede and the Wyverns followed the line of the Twente Canal. The 5th Dorsets and 1st Hampshires captured Hengelo on 3 April, where they found a factory which made the latest type of AA predictor. The Canloan officer, Lt W. I. Smith with 4th Wiltshires, tried to get his platoon to sleep in the loft of a barn:

> It was impossible. The Dutch [farming] family were overjoyed at being liberated. There were two or three nuns in the house. When I returned from Company HQ I found the Dutch – boys, girls, even old men and women as well as the nuns with British troops of all ranks hand in hand dancing in a ring and singing Dutch and English songs. I have rarely seen such pure delight as was shown by these Dutch people.

From Winterswijk to Lochem near the Berkel River and canal, the Wyverns met strong resistance from the 23rd SS Freiwilligen Panzergrenadier Division (Nederland). Unfortunately there were well over 20,000 young Dutch SS troops who, when captured, faced an uncertain future. In Borne, 3 miles ahead of Lochem, the Dorsets were so popular that the main square was renamed Dorset Square!

In Bawinkel, AVRE petards had to be fired to destroy roadblocks and Captain J. A. H. Clark, 7th Somerset Light Infantry, remembered the battle of Buckelte-Hamm: 'on a very hot fine day. Off we went, Kangaroos, tanks, TCVs [troop-carrying vehicles], appendages of every sort ... so we motored into Buckelte, a charming village in the sunshine – daffodils, forsythia, tulips, jonquils, fat farms and blossom. All the young Nazi troops were sunbathing and captured; dozens were caught hastily changing into civvies [civilian clothes].' In Haselunne two gin distilleries were captured and Colonel Alabaster, the 30th Corps Welfare Officer, rushed up to take over the stocks for the NAAFI to sell! Another centre line towards Cloppenburg was christened 'Heart Route'. Holte and Herzlake were taken and by dawn on 11 April, 4th Dorsets in Loningen outfought a battalion of the Grossdeutschland Brigade of young cadets. Herssom and Vinner fell to 4th Wiltshires but at Wachtum mines and a huge road block held them up. Pte R. Gladman, 4th Wiltshires, noted: 'We carried on through Germany, sometimes walking, marching, sometimes

Royal Norfolks with 'Big friends', on way to Bremen.

in a TCV. If the road was blocked by fallen trees, vehicles would have
to go round through the grass. To avoid landmines, we crawled in line
probing with bayonets but a problem with anti-personnel mines. Our
officer said to step on one would mean the loss of your wedding tackle, a
cause of problems on your honeymoon.' Cloppenburg took two days for
7th Hampshires 5th DCLI to capture, which resulted in 300 cadets being
killed, wounded or captured. Inevitably, in street fighting at night, the
Wyverns suffered badly. On 14 April, the river Leithe was reached near
the Ahlhorn crossroads. In the dense, dark and gloomy spruce and pine
forest of Cloppenburg, Major Peter Hall of the 1st Worcesters noted:

> The vanguard company's job was frustrating and hazardous hard work. Along
> this single axis they had to contend with mines, road blocks, booby traps and
> periodic mortaring and occasional sniping for good measure. They reached
> the river Leithe, 20 yards wide, with steep banks and the bridge was blown.
> On the 15th about 0545hrs there was a tremendous artillery thunder on my
> forward platoon, followed by a massive enemy infantry and tank assault.
> From my Company HQ I saw the forward platoon soldiers running back
> hastily. I organised the new forward positions. My two signallers were killed.

The wireless was destroyed. The only link was through the regimental wireless network of 15/18th Hussars. We were on our own. For the next two hours the battle raged through the woods. This was one of the last serious counter-attacks of the war. We took numerous prisoners, between 400 and 600 enemy dead and one Royal Tiger tank out of action. The Germans included Luftwaffe, panzer units, Waffen SS and Wehrmacht troops.

Both the Sherwood Rangers Yeomanry and 13/18th Hussar Sherman tanks gave invaluable support to the Wyverns in the capture of Cloppenburg and in the counter-attack in the forest.

Most historians have glossed over the number of quite extraordinary, bitter and determined defences made, usually by the Hitler Jugend, by cadets of all ages, and always by the Waffen SS. It has been seen how the Teutoburger Wald battle sucked in some of Montgomery's troops. There are some more examples.

It took the 15th Scottish Division five days – all of the division with massive supporting arms – to take the towns of Uelzen and Stadensen. The first was a key communications centre as part of the German defences west of the river Elbe. Fresh troops had just arrived from Denmark, including reserves of panzergrenadiers, and a Flak regiment with 20mm cannon mounted on trucks. They were fresh, well-fed, truculent, undefeated and longing for a fight. There were also youngsters

Seaforth Highlanders clearing houses under fire in Uelzen, 18 April 1945.

Tac HQ 22 Armoured Brigade, Brigadier Wingfield and his staff.

from the GAF, the Luftwaffe ground troops. At dawn on 14 April the Highland Light Infantry were devastated around Holdenstedt two miles south-west of Uelzen, which has the little river Ilmenau flowing through it. Panzer Division Clausewitz overwhelmed the 15th Scottish Division Recce Regiment at midnight on 14/15th, causing sixty casualties and capturing thirty-nine prisoners. Some of the attack force were very drunk French SS troops, who howled, screamed and took outrageous risks in the village of Nettelkamp. In Stadensen, the Glasgow Highlanders were attacked at four in the morning. Most of the houses were set on fire. Scottish ammunition trucks blew up and the dreadful ordeal went on until dawn. The Scots had fifty-four casualties including twenty-two taken prisoner. Scores of German civilians were also killed during the night. On 15 April the hapless HLI were attacked by a Light AA group from Denmark with SP guns, bazooka squads and plenty of Spandaus and they, and the Gordons, were pinned down by fire all day. By the 19th, Uelzen was in ruins, bloodstained ruins.

The Guards Armoured encountered 2nd Marine Division and their GOC, Major General Adair, wrote: 'They were nearly all ex-sailors, lately submarine crew with little time for military training, lacked the fighting skill of the paratroops but their discipline and bravery were exemplary.

241

PBI on their 'Big Friends'. Better than walking. (*Imperial War Museum*)

They were well equipped with bazookas, used in daring fashion. Their tactics often involved them in annihilation, but their aggressive spirit certainly delayed our progress most effectively.'

The Scots-Welsh battle group attacked Visselhovede from the south and the Coldstreams attacked from the north-east. 'A furious battle arose suddenly without any warning. Our mortar carriers were destroyed by bazookas; a bombardment of the houses where the battalion HQs were sited. Fighting spread along both sides of the main street. For two hours the situation was hopelessly confused.' Fighting went on into the evening. Eventually the German Colonel Jordan and 400 marines were captured from 5th and 7th Marine Regiments. The Guards Armoured Division then took Tewel, Deepen, Sohlingen, Scheesel (a hospital town with 2,000 German wounded), Rotenburg (another important hospital centre), Heidenau and Sittensen. Most of these villages were briefly defended but in Wistedt and Elsdorf there was serious trouble. Lieutenant von Buelow, of 15th Panzergrenadier Regiment destroyed the Irish guardsmen and recaptured Wistedt. Guardsman Eddie Charlton, the only survivor, manned his Bren gun valiantly, but was hit many times and was killed. His heroism earned him a posthumous Victoria Cross. Elsdorf's Irish defenders were badly mauled by the 15th Panzergrenadiers.

In their last battle of the war on 'Black Friday', 13 April, the 1st Hampshires of the Wessex Wyverns had a bitter battle for Cloppenburg against the officer-cadets of the Grossdeutschland.

The 53rd Welsh Division found Battlegroup Hornemann and the 5th Marine Regiment defending Rethem and its bridge over the river Aller. The German 2nd Marine Division had obviously taken Himmler's death penalty threat rather seriously. The three grenadier regiments were backed by an artillery regiment, an anti-tank (88mm) regiment, engineers and fusiliers. They were all naval personnel aged 17–23 and their GOC, General Hartmann, an ex-U-boat commander put up a ferocious defence, unfortunately for the Welsh. There were also experienced SS in Rethem, so it took the Welshmen nearly three days to succeed.

The highly experienced Brigadier H. Essame, OC 214th Brigade, noted: 'It must be admitted that Lt General Erdmann conducted the withdrawal of the German armour to Bremen with great professional skill. Travelling in a captured jeep he raked together the remnants of many formations into battle groups which still fought on.'

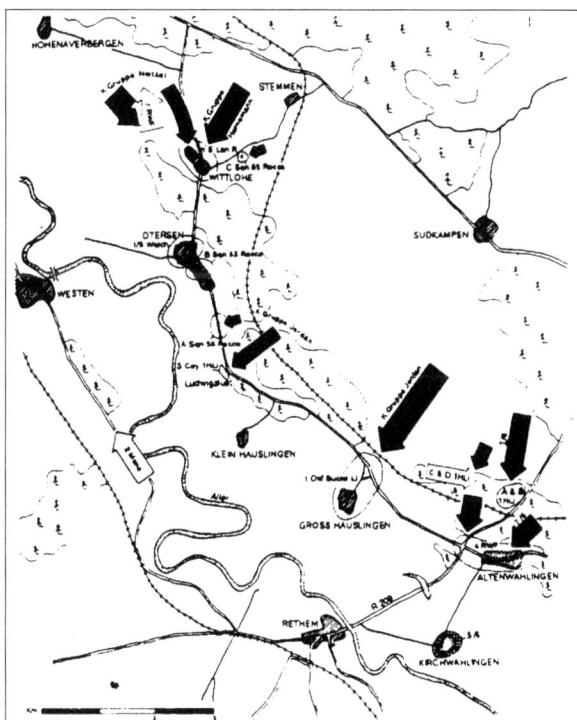

The savage battle for Rethem.

CHAPTER 19

Capture of Bremen: Operation Clipper

'Bremen lies astride the river Weser, the greater part of the town is on the north of the river. To the south the town is only some two thousand yards before the fields begin. But outside this there are a series of villages or small towns which form a suburban ring. On the west there is Delmenhorst [a significant smallish town], Brinkum, Leeste, Arsten and Dreye which was only 300 yards from the river Weser. The fields between these villages were very low and had been extensively flooded by the enemy. North of Brinkum was flooded to a depth of four feet. Between Sudweyhe and Dreye there was little flooding but the ground was so soft that tanks could only move on roads. A wide stream flowed across the front from the floods north of Brinkum between Sudweyhe and Dreye to the Weser.' This was Major Guy Radcliffe's, 2 i/c of 2nd KSLI, description of the battlefields of the six-day Operation Bremen.

Bremen was the tenth largest city-port in Germany, with a population of well over half a million. Although it is 37 miles upriver from the North Sea, it is bisected by the great river Weser, due south of the main port of Bremerhaven. It was also very important to the Allies. Antwerp was the major port supplying the British and Canadian forces, and Bremen and Bremerhaven had been 'reserved' for General Eisenhower's armies, who were still being supplied from the Mediterranean and Normandy, mainly Cherbourg because Brest was still being repaired.

'Bomber' Harris had been blasting Bremen with saturation bomb attacks since 1942. The main targets were the Focke-Wulf aircraft factory, Atlaswerke shipbuilding works, Bremen Vulkan shipyard, Deschlwag (AG Weser) shipyard, Korff oil refinery, Norddeutsche Hutte steel works, Valentin submarine pens, Borgward motor transport plants and Bremen

One of the heavy AA guns knocked out by five 5 RTR tanks at Rethem.

Osleghausen railway station. There was an airfield just south-west of the town and two concentration camps (obviously not targets) at Bremen-Farge and Bremen-Vegesack. Nor of course was Becks' famous brewery be a target. Both Adolf Hitler by radio, and Himmler by paying a visit, had ensured that Bremen was designated a *festung*, a fortress, and in theory to be defended to the last man.

The remains of the 1st Parachute Army, which had been battered during operations Veritable, Blockbuster, Plunder and many small, nasty actions east of the Rhine, had now reinforced the Bremen garrison.

Lt General Fritz Becker was in overall command of the 40,000 defenders of Bremen and its ring of suburban defences. Major General Sibert, a big fat man with a broad, unintelligent sour-looking face, and Colonel Mueller had the powerful 8th Flak Division under command. Its dual purpose 88mm AA guns had been trying to ward off Bomber Harris's RAF bombardments. The 280th Wehrmacht 'Stomach' Regiment, 6,000 Volksturm 'veterans', and a 2,000-strong mixed bag of Ost, coastguards, police, and including 500 first-class soldiers from 18th SS Horstwessel Training Regiment, were all part of Becker's command. There were also over 2,000 Kriegsmarines, either unemployed U-boat crews, marine cadets, or other ship-less naval bodies commanded by a vice-admiral. This immense flotsam and jetsam, of paratroops, SS Flak troops, Ost, Wehrmacht, Volksturm, police and fire brigade would fight hard against overwhelming odds in the last major battle fought by Montgomery's British Liberation Army in the Second World War. It was going to be a six-day battle. Operation Cricket, against the cream of the infantry divisions, included: Whistler's 3rd British, Thomas's 43rd Wessex Wyverns, Hakewell-Smith's 52nd Lowland and MacMillan's 51st Highland. In support were Brigadier Carver's 4th Black Rat Armoured Brigade and a vast array of 'Funny' flame-throwers, Crocodiles, AVRE bombards, some Flails and the *schwim-panzers*, which would cause consternation.

Lt General Brian Horrocks consulted Montgomery about the complicated plans for the much-increased 30th Corps to capture such a large city surrounded by flooded fields, astride a major river.

Intelligence had shown that the powerful 15th Panzergrenadier Division, a first-class formation, had moved from the west of Bremen to the east, to deny the road and rail links between Bremen and Hamburg to the British. The implication was that Bremen would be strongly defended. Nevertheless Horrocks tried for a peaceful move. Nearing the end of the war he certainly wanted more than ever to minimise casualties. By 20 April 3rd British and 51st Highland, two key 'chess pieces', were in position, south-west and east respectively, of Bremen. So propaganda leaflets, in German naturally, were fired by 33rd Field Regiment RA in 500 special non-lethal shells over Bremen: 'The choice is yours! The British Army is lying outside Bremen, supported by the RAF and is about to capture the city. There are two ways this can take place. Either by the employment of all means at the disposal of the Army and the RAF or by the occupation of the town after unconditional surrender … You have 24 hours to decide.'

However, General Becker was never going to surrender tamely. Bomber Harris unleashed 500 Lancasters and Halifaxes flying at 4,000 feet on the afternoon of Sunday 22 April, and continued to send them in day after day until the ground forces in the smoking ruins were satisfied. The ubiquitous Chester Wilmot was there:

23 April 1945. The telephone they used was found in the station at Achim, ten miles east of Bremen, when the 52nd Lowland Division took Achim this morning. This afternoon I was there when the ultimatum was phoned through. The stationmaster at Achim was asked to call up the stationmaster

Infantry of 156th Brigade 52nd Lowland division and Sherman tanks of the 4th Armoured Brigade entering Bremen.

REs support the 52nd Lowland Division clearing roadblocks, Bremen, 26 April 1945. (*Birkin Haward*)

52nd Division clearing the ruins of Bremen.

at Bremen, and then he handed over the phone to our intelligence officer. He spoke to a German lieutenant, who then went off to convey the message to the Bremen garrison commander, and to bring back a senior officer to discuss the ultimatum. We waited for over an hour. Then the phone from Bremen rang again. Not to tell us the reply, but to give us the friendly warning that the RAF was bombing Bremen, and to say that no senior German officer would come to the phone. Then the line went dead. The bombs had evidently cut it.

We couldn't see much of the bombing because of the low cloud, but hundreds of Lancasters attacked the ack-ack defences and installations in and around the city. We've warned Bremen that yesterday's heavy raid is only a small foretaste of what is to come if they try to hold out, and already our heavy and medium guns are pounding away at the defences. Occasionally we're putting over a few more propaganda shells, filled with 'free-passage' leaflets, for those who've had enough. And these appeals are being reinforced by loudspeaker vans blaring out across no-man's land on the fringe of the city; but so far the Germans have shown no signs of cracking.

No doubt, when Himmler visited Bremen in his special armoured train last week he tightened up the Nazi grip on this port. A Nazi chief has been installed

Glasgow Highlanders fighting in the Focke Wulf Factory at Bremen.

to organise control of the SS, the Gestapo, and the local police, and to make sure that the garrison commander, General Becker, doesn't weaken. Civilians who've escaped say that already the local führer has sacked the burgomaster, who wished to surrender. Nevertheless the plight of the city is now serious; it's three-quarters encircled and the Germans control only two good roads leading out of it, and these only lead north into the peninsula which we've already sealed off between Bremen and Hamburg. Refugees say that inside Bremen there are about 300,000 German civilians; some 50,000 foreign workers and more than 20,000 troops are around the city. Civilians also say that they've been short of food in recent weeks, but their greatest problem now is going to be shortage of water. The city's local supply has been put out of action by bombing; and their main outside source, a pipeline from the Harz Mountains, was cut off today by us just before the RAF bombing began.

Today they also lost their main source of fuel, an oil refinery at Achim which the 52nd Division captured this morning. The refinery and its storage tanks are hidden underground in a wood near Achim, and it was working until yesterday. In fact petrol and oil from this refinery moved the 15th Panzergrenadier Division when it transferred from the south to the north of Bremen last week to counter the Guards' outflanking movement. There are, no doubt, other storage tanks in Bremen, but the loss of this refinery at this stage must be a considerable embarrassment.

The Germans are now being driven steadily back into Bremen and the garrison of 20,000 may well be increased to 30,000 by the time all the units fighting outside the perimeter have withdrawn inside. There they'll have a very strong defensive position, for Bremen is very difficult to attack. It stands almost on an island in the flood plain of the Weser. On the south it's protected by a belt two miles wide which the Germans have flooded, and on the north-east and north the low-lying ground is laced with dikes and canals.

Borstel, two miles north-east of Achim, cost 5th HLI fifty-one casualties but yielded 242 prisoners and thirty-nine guns from 88mm to 20mm. 5th KOSB pressed on to Embsen, a mile ahead.

The ring of strongpoints around the city had to be laboriously overcome. 3rd British were tasked with capturing Brinkum, which was defended by four companies of young SS fanatics of the Horstwessel Battalion. Supported by Crocodiles from 7th RTR, Shermans of 4/7th Dragoon Guards, and a troop of Valentine SP 17-pdrs, it took the Suffolks, Norfolks and Warwicks almost two days to succeed. Field Marshal Montgomery visited the division on the 20th and Major General 'Bolo' Whistler wrote in his diary: '20 April. Monty keeps a very close eye on everything. He has certainly done me jolly well and made my battle both possible and profitable. We have captured five officers and 1,259 men on April 15th and another 1,800 between 15–19th. We have killed about 200–300 more. Not much shelling from the Boche for which I

German soldiers, some only young boys, surrender to the British in Bremen, 26 April 1945.

One of the young German soldiers surrendering to the British, 26 April 1945.

Blasted, ruined Bremen, 26 April 1945. (*Birkin Haward*)

am thankful. Lovely weather! Had a close look at Bremen yesterday. It appears rather undamaged in spite of 'Bomber' Harris and his efforts.'

The 51st Highland Division had captured Rees in Operation Plunder and had moved north via Empel to Isselburg and Anholt. Every infantry unit in the British Army desperately needed reinforcements after ten months of intensive fighting. The 1st Black Watch stayed in Isselburg until 5 April and received reinforcement officers from LAA, the Suffolks,

Argylls, Duke of Wellingtons, West Yorkshires, Durham Light Infantry and redoubtable Canloans. 'They all proved excellent Black Watch officers as if to the manner born,' wrote Lt Colonel John McGregor in *The Spirit of Angus.*

Two unusual features appeared now in every advance east towards Bremen, Hamburg and the Baltic. In addition to the inevitable panzerfaust teams, which caused many tank casualties, every road was 'seeded' with huge sea-mines – the large black ball filled with HE and with little pressure-spikes. Buried to a depth of three feet they blew everything that set them off to kingdom come. Comets, Shermans, half-tracks and carriers were simply blown to pieces. It was Hitler's unemployed naval Kriegsmarines who were responsible. The other strange feature was the plethora of German hospitals crammed full of wounded. The Highland Division found one in Vechta with 2,500 patients, one in Lohne with 250 and three more around Quakenburg. When the Highland Division reached Delmenhorst, which was a large, undefended town that had been reached by the Derbyshire Yeomanry recce armoured cars on 20 April, the burgomeister of the town informed them that there were no fewer than fifteen hospitals in the town crammed full with German wounded.

Major Martin Lindsay, second in command and often in charge of the 1st Gordons, calculated that up to 27 March, 102 officers had served with his battalion. The usual strength was thirty. The average service of the fifty-five officers commanding the twelve rifle platoons was thirty-eight days or five and a half weeks. Just over half had been wounded, a quarter killed, 15 per cent invalided home and 5 per cent had survived. These terrible figures were duplicated in most of the infantry divisions. 'Everybody had a prejudice against being killed in the last month of a six-year war, so people were playing for safety ...'

The Highland divisional 'cage' harboured nearly 200 men on charges of desertion. Many were veterans of North Africa, Sicily, Italy or Normandy who, poor devils, had exhausted their store of courage. Many others were eighteen-year-old reinforcements, poorly trained and disciplined, who were just completely out of their depth. A third group were from rear lines of communication that had been hauled out of comfortable safe jobs and thrown into the cauldron of the 'sharp end'. It seemed to be lunacy that the SS, the Hitler Jugend, the officer cadets, the paratroopers and the young marines were all prepared to fight, fight and die.

German prisoners in Bremen.

George Blake's history of the 52nd Lowland Division notes:

> On April 12 the troops knew that their next objective was Bremen, the
> glittering prize of victory, some 30 miles ahead. Now they were to enjoy the
> pleasures and triumphs of 'swanning' ... to go 'swanning' is to ride in or on a
> vehicle through strange country with little interference from the enemy. For
> an *infantry* division in particular to go swanning encouraged the gala mood,
> so that the troops, true to the British code of humour, rode on tanks [Scots
> Greys to Etelsen, 44 RTR to Volkersen, with 2 KRRC protecting the right flank
> from Langwedel to Bassen and Baden on the Bremen-Hamburg autobahn], in
> strange headgear, waved to the native children, chirruped facetiously at the
> womenfolk of the countryside and generally took Germany for their own.

It was the 6th Cameronians who took Etelsen and Baden, and 4/5th
Royal Scots Fusiliers who finished off Baden, but 5th Highland Light
Infantry were counter-attacked during the night of 21/22 April. Achim
was an important road junction captured by 7/9th Royal Scots and
4th KOSB. In three villages the unfortunate ersatz 'ear and stomach'
defenders and Flak troops had 750 put 'in the bag'.

Just before the move on Bremen, the fiery GOC of the 43rd Wessex
Wyverns was not a happy man. 130th Brigade was loaned to 52nd

253

In Bremen the fire brigade was captured, and is here seen lining up for checking.

Lowland to provide fire support, 129th Brigade was lent to 51st Highland and 214th Brigade was left behind to mop up German pockets of resistance around Ahlhorn, so for a short time Thomas was a general without a division.

The 30th Corps' plan was for 51st Highland to make a feint attack on Delmenhorst on the west side of Bremen, which in the event was surrendered by the burgomeister for humanitarian reasons. The 3rd British in Buffaloes would put in a surprise attack across flooded countryside from the south-west, while 52nd and 43rd attacked the main part of the city from the south-east.

The Crocodile flame-throwers of 7th RTR were much in demand and were involved in actions in Uphusen, Mahndorf and Hastedt. B Squadron helped 52nd Lowland to capture 800 defenders on 25 and 26 April, and A Squadron flamed 3rd British into Dreye, Arsten and Harlenhausen. The AVRE petards of 16th, 82nd and 22nd Assault Squadrons blasted their way into Bremen clearing road blocks, laying a skid Bailey bridge and dropping fascines into the many huge craters caused by the sea-mine explosives. Wynford Vaughan-Thomas wrote for his BBC programme:

157th Infantry Brigade 52 Lowland division clearing Bremen docks.

26 April 1945. I am in the centre of the city of Bremen, if you can call this chaotic rubbish-heap in which our bulldozers are working as I speak, their men with handkerchiefs round their throats to avoid the flying dust, if you can call this rubbish-heap a city any more. There are walls standing, there are factory chimneys here and there, but there's no shape and no order and certainly no hope for this shell of a city that was once called Bremen. It's the sheer size of the smashed area that overwhelms you. These endless vistas down small streets to the houses leaning at drunken angles and this inhuman landscape of great blocks of flats with their sides ripped open, and the intimate household goods just blown out into the bomb craters. Bremen was a city of many thousand inhabitants, but once you are past the outer suburbs there is hardly a house intact in which these people could live, because they haven't even tried to live in them. Everybody has gone into those giant air-raid shelters, these great blocks of almost solid concrete that are dotted all over the ruins of Bremen. They stood up to the bombings well, and now, from the one opposite us, the people are coming out, to stare at the tanks and the bulldozers, and at the guns now rolling past us into the heart of the city. Some of these people are already starting to potter about in that seemingly aimless way that bombed-out people do among the dusty rubble.

Some of the fifty submarines captured by 52nd Division on the stocks at Bremen.

Burgher Park, a junction of six roads, also housed fourteen underground bunkers. On the way there, Captain C. N. B. Hancock, 94th Field Regiment RA, wrote: 'I was FOO with the Somersets. Darkness had fallen and the Crocodiles were put to work on a couple of large houses which were giving a lot of trouble. It was a grand sight to watch them belching great tongues of flame, 70 or 80 yards long and in a short time both houses were burning furiously. The Boche inside had an unpleasant choice, the flames within or the waiting machine guns of our infantry and tanks outside. The battle was now well under control with the tanks, Crocodiles and infantry providing mutual covering fire as together they assaulted house after house.'

The 3rd British meanwhile had to capture the bridge over the Kattenturm Canal intact in order to clear the aerodrome and Focke Wulf aircraft works south of the river. Two journalists, one from *Reuters*, the other from Associated Press, watched the extraordinary 'marine' attack as a fleet of forty-seven Buffaloes of 4th RTR under Captain Harris made 115 trips across three feet of flood water to carry 2nd Royal Ulster Rifles over 2,000 yards of flooded fields to capture Arsten. Across

General Horrocks talking to a group of soldiers after the capture of Bremen, April 1945.

the bund they went and captured the vital bridge before it could be blown. The sappers made safe all the huge bombs around the bridge. Later 1st KOSB and 2nd Lincolns occupied the aerodrome and factory, capturing 350 prisoners. The German general later said: 'It was not fair using 'schwim-panzers' like that!' The 3rd British pushed steadily into the Neuenlander suburbs and into the barracks. Iain Wilson, 1st KOSB wrote: 'In the chilly morning light of the 25th, as the guns continued to pour shells into Bremen, one could feel that the great port was in its death-throes and that its doom was approaching. The approaches to the city were flooded and in the ruins we knew the SS were waiting to fight it out. Life became more precious and desirable as the chances of death or wounds became less.'

Bomber Command kept up their relentless attack on Bremen for 5 hours continuously on the 24th, by which time 'Von Thoma', as Thomas was nicknamed, had regained control of his lost sheep. Brigadier J. O. E. Vandeleur (he of Guards Armoured fame during Market Garden) had been promoted in rank and joined the Wyverns. He wrote:

The attack upon Bremen was the last big battle in which we were engaged. My 129th Brigade was to capture the Burgher Park, a stronghold on the eastern side of the city. 52nd Lowland were to capture the U-boat yards. We approached Bremen cautiously two up: right 4th Wilts, left 5th Wilts. I kept the Somerset LI in reserve. Our opponents were German marines who were lusty fighters. I had been instructed not only to capture the Burgher Park but also the army, naval and Luftwaffe commanders. When darkness fell I decided to launch 4th Somersets under Colonel Lipscomb and smash our way into the big bunkers where the German HQ and core of their defences were situated.

From Rockwinkel the two Wiltshire battalions advanced steadily, leap-frogging, platoon by platoon, company by company. A huge seven-storey-high Flak tower was taken, and its garrison of 160 surrendered without a shot fired. AVREs petarded each roadblock and Crocodiles flamed difficult targets and the Sherwood Ranger Shermans did their stuff.

Brigadier Walter Kempster, OC 9th Brigade 3rd British, noted on the 26 April: 'At first light I passed 2 RUR through 1 KOSB to capture the gasworks and docks on the south bank of the river. By then the Hun was utterly demoralised. 2 RUR just walked on to their objectives and by 0930 I reported to Division that we'd done our job – nearly 30 hours in advance of the estimate.' Some interesting spectacles were seen – the bag of prisoners included, besides many SS, some weary old men and young boys, a sadistic displaced person camp commandant and a large portion of the Bremen police force. Kempster's men also took prisoner 800 from the 18th SS Regiment and Hitler Jugend, whose average age was sixteen–seventeen, the Chief ARP Warden of Bremen, a city fire engine crew and many Volksturm, elderly chaps complete with well-packed suitcases. At daybreak on the 27th, the 4th Wiltshires of the Wessex Wyverns battered their way into the main bunkers. Major Pope captured there Lt General Fritz Becker, a vice-admiral, the Nazi Bishop of Bremen, Major General Siber, and thirty other senior officers. Within 36 hours the Wiltshires had taken two generals and over 600 other prisoners. And a considerable amount of Zeiss binoculars! Brigadier Essame wrote: 'All day on the 27th prisoners streamed into the Divisional cage. By nightfall 2,771 including 94 officers and 831 hospital cases. The people were broken spirited and lifeless, docile, bewildered and hopeless. Fighting, rape and open murder broke out among the thousands of released slave labourers. We handed over this charnel house, once a civilised city to 52nd Lowland Division and moved out into the clean air of the Cuxhaven peninsula on the 28th.'

The Lowlanders had done well. The Glasgow Highlanders took their two factory objectives. The 6th Cameronians seized the whole suburb of Hemelingen and 4/5th Royal Scots Fusiliers found themselves in the

The battle of Bremen.

heart of Bremen. 186th Field Regiment RA shrewdly took over the great sports stadium on the south side of the city for its gun positions. And the 7th Cameronians and 54th Anti-tank Regiment, in its recent new role of an assault SP unit, came into the dock area to check on one-man submarines. A German full admiral in charge of the dock area insisted on a formal ceremony of surrender of his motley marines to 5th KOSB. They included four Helferin (German WRENs). The burgomeister of Bremen, not to be outdone, had a parade of a thousand very mixed soldiers, and insisted on a series of barrack square drill movements before surrendering to a slightly bemused 4th KOSB. The Lowland historian, George Blake, who was present, wrote: 'The control of this bear-garden was a heavy responsibility on the cool and terribly tired Scottish soldiers who had taken the city after such a long and gruelling battle. It was immensely complicated by the litter of smashed and fallen masonry left by the bombers. A central area of the city, some four miles by one, had ceased to exist.'

The 'Thieving Magpies' T-Force and 30th Assault Unit sent separate foraging units ahead into Bremen. For T-Force, 5th Kings and the

Buckinghams seized the city's main police station as their HQ, from which they could conduct a week's detailed search of the city. There was an American interest as well, and 125 targets had been previously identified. Lt Commander Dalzel-Job sent his engineers out to work. They blew ninety-five safes, which contained useful documents and literally billions of reichsmarks. The huge bunker-shipyard at Bremen-Farge – a concrete monster – housed thirteen pens for 150 of the ultra new high-speed U-boats. At the Descimag U-boat assembly plant a Narvik class destroyer and sixteen U-boats nearing completion were found. T-Force put a guard over the Focke-Wulf aircraft engine factory. these were all key places and records for the incoming American HQ when it was set up.

Major Raymond Burt, 22nd Dragoons, whose Shermans (without Flails) had been supporting 2nd RUR, 2nd Lincolns and 1st KOSB for the six days of fighting, described the last days of Bremen:

> So this was how the city was falling – without fight, in rain and betrayed by those who had brought it to its present squalor. For all their boasts and threats, the Nazi leaders had gone and the city was abandoned to a few thousand AA gunners and marines and the old men and women and children who waited our arrival in the air raid shelters. The advances into the suburbs were something of a formality. But they were carried out, block by block, with care and precision – companies and supporting tanks leapfrogging through one another along the silent and mined streets. It looked formidable enough. The road blocks were defensible. Slit-trenches and anti tank ditches had been dug across street intersections. Enormous land-mines had been laid and wired ready for explosion by the roadsides. The windows of the ruined houses provided the 'heroic twilight' of the Nazis of the city of Bremen. But no shot was fired. Hitler's war was running down – it had collapsed into a dreary mopping up operation which went on because no one in authority had the desire to cry 'stop'.

CHAPTER 20

Alpenfestung: Göbbels' Hoax

In the autumn of 1944 stories began to circulate in Switzerland that the Nazis had set up a vast military complex in the mountains of southern Germany and Austria. They probably derived from Dr Josef Göbbels, that master of spin. Agents of the SD (SS Intelligence Department) forwarded this 'information' to Gauleiter Hofer of the Tyrol. In November, Hofer sent a report to Martin Bormann, the deputy Führer, to the effect that US experts had forecast unacceptably high casualty rates for the campaign by the Americans to overcome this final bastion. Moreover Hofer noted that 'East-West tension will become visible if the war drags on too long'. This situation could be exploited by the Nazis who should open negotiations with the West at the expense of the Soviet Union. Hofer urged that a defence redoubt *festung* be set up in mountainous areas, such as his own Gau (shire) in the Tyrol. In fact, Hofer had strengthened areas already, which he modestly called 'The Hofer Line'. He recommended that Allied POW camps be moved there to deter Allied air interdiction. The garrison would need élite military units backed by 30,000 men of the Standschutzen (local defence unit's equivalent to the Volkssturm). General Ritter von Hengl was ordered by Berlin to command the redoubt garrison, but his mountain troops' corps, between 1–5 April 1945, consisted of only 30,000 men. He produced a thoroughly damning report and dismissed the Hofer plan as a slogan: 'The Alpine Redoubt existed merely on paper.'

Göbbels wrote about Hofer's recommendations as though they were operational: food stocks were already stockpiled; underground factories were in full production; the élite units garrisoning the *Alpenfestung* were in action; and German scientists were working hard to produce another super vengeance weapon, which would ensure a German victory!

261

On 11 March 1945 SHAEF produced this intelligence summary: 'Accumulated information and photographic evidence made possible a more definite estimate of the progress of plans for the "Last Ditch Stand". German defence policy is to safeguard the Alpine Zone. Defences continue to be constructed in depth in the south through the Black Forest to Lake Constance and from the Hungarian frontier to the west of Graz … In Italy defence lines are built up in the foothills of the Italian Alps. The results of air reconnaissance show at least 20 sites of recent underground activity as well as numerous natural caves mainly in the region Golling, Feldkirch, Kufstein and Berchtesgaden [Hitler's personal chalet-retreat atop a mountain] … accommodation for stores and personnel. The existence of several underground factories has also been confirmed … As regards actual numbers of troops, stores and weapons … evidence indicates considerable numbers of SS and specially chosen units are engaged on some type of defence activity at the most vital strategic points … The most important ministries are established within the Redoubt area … while Hitler, Göring, Himmler and other notables are said to be in the process of withdrawing to the mountain stronghold.'

At SHAEF a large map was on display headed 'Unconfirmed Installations in Reported Redoubt Area' covering 20,000 square miles, the Alpine meeting point of southern Germany, northern Italy and western Austria with Hitler's Berghof near Berchtesgaden in their midst. Moreover symbols were printed on the base of the map indicating ammunition and fuel dumps, HQs and barracks, food depots, radio installations, factories, troop locations and a chemical warfare site.

Adolf Hitler received Hofer's November 1944 memorandum during April 1945 and ordered that the gauleiter's plans be put into immediate effect. At that late stage it was an impossible pipedream.

Nevertheless Eisenhower, Bradley and Lt General Walter Bedell Smith, in particular, felt strongly that: 'After the Ruhr was taken we were convinced there would be no surrender at all so long as Hitler lived. Our feeling then was that we should be forced to destroy the remnants of the German Army piece by piece, with the final possibility of a prolonged campaign in the rugged Alpine area of Western Austria and Southern Bavaria known as the National Redoubt.'

The SHAEF summary continued: 'This area is by the nature of the terrain practically impenetrable. The evidence suggests that considerable numbers of SS … are being systematically withdrawn to Austria …

American paratroops reached the Berghof, Hitler's retreat in the Bavarian Alps, in May 1945. This is Hitler's own residence, with an RAF bomb crater in the foreground.

The Berghof, April 1945. An American GI looks at Hitler's home. The Nazis' 'last stand' in a non-existent Alpine redoubt.

Here defended both by nature and by the most efficient secret weapons yet invented the powers that have hitherto guided Germany will survive to reorganise her resurrection. Here armaments will be manufactured in bomb-proof factories, food and equipment will be stored in vast underground caverns and a specially selected corps of young men will be trained in guerilla warfare [Werewolves] so that a whole underground army can be fitted and directed to liberate Germany from the occupying forces.'

General Eisenhower's report 'D-Day to VE Day' published in 1946 devotes several pages to the Redoubt. There is no doubt that Eisenhower concluded that the National Redoubt was now of greater significance than Berlin: 'Military factors when the enemy was on the brink of final defeat were more important in my eyes than the political considerations involved in an Allied capture of the capital since this no longer represented a military objective of major importance.' By taking a unilateral decision (but backed by General Marshal in Washington) his new plans involved his right wing of armies driving south-east to meet the Russians in the Danube valley, west of Vienna, and *seizing the Redoubt before the Nazis could organise it for defence.*

Churchill was furious. Montgomery was furious. They had always wanted the Allies to get to Berlin come what may. On 28 March Eisenhower sent his plan direct to Marshal Stalin, who was undoubtedly delighted. Ike wrote: 'The Prime Minister was greatly disappointed and disturbed because my plan did not first throw Montgomery forward with all the strength I could give him *from American forces* in a desperate attempt to capture Berlin before the Russians.'

Alpenfestung was of course a brilliant Göbbels hoax! What actually happened was that by mid-April most of the key ministries from Berlin had been moved to the Berchtesgaden area, and had taken their documents and administrators with them. Most of the OKW staff (Supreme Command) had also been transferred to the Redoubt area. However, on 20 April, Hitler decided to stay in Berlin – and die there – defending the city to the last. SHAEF had probably noted the movement of staff to the area – and assumed that the Wagnerian Götterdammerung would follow. A new German Army, the 11th, under General Walter Wenck, had five modest divisions which, in theory, guarded the Redoubt as part of their command area. On 18 April Wenck's 10,000 men were trapped by the US 1st Army around Dessau near the Harz Mountains. That was the end of *Alpenfestung*.

Bibliography

BBC War Report, Despatches: D-Day to VE Day (OUP, 1946)

Blake, George, *Mountain & Flood* (Jackson & Co, 1950)

Churchill, Winston, *Triumph & Tragedy* (Cassell, 1951)

Cornwell, John, *Hitler's Scientists* (Penguin, 2004)

Davies, Norman, *Europe at War* (Pan, 2007)

De Guingand, Francis, *Operation Victory* (Hodder & Stoughton, 1941)

Delaforce, Patrick, *Churchill's Secret Weapons* (Pen & Sword, 2006)

Delaforce, Patrick, *The Rhine Endeavour* (**Amberley, 2010**)

Ford, Ken, *Rhineland 1945* (Osprey, 2000)

Gilbert, Martin, *The Day the War Ended* (Harper Collins, 2001)

Gilbert, Martin, *Second World War* (Phoenix, 2009)

Hastings, Max, *Finest Years (Churchill)* (Harper Press, 2009)

Hastings, Max, *Armageddon* (Pan, 2004)

Hauge, Jens Chr, *Liberation of Norway* (Norsk Forlag, 1950)

Hitchcock, William, *Liberation, Bitter Road to Freedom* (Faber, 2009)

HMSO, *By Air to Battle (Airborne Divisions)* (1945)

HMSO, *World War 2 Collection* (2001)

Horrocks, Brian, *A Full Life* (Collins, 1960)

Longden, Shaun, *T-Force, Race for Nazi Secrets* (Constable, 2009)

Lucas, James, *Last Days of the Reich* (Arms & Armour, 1986)

Moorehead, Alan, *Eclipse* (Hamish Hamilton, 1945)

Petrow, Richard, *The Bitter Years* (Purnell Book Services, 1975)

Rosse & Hill, *Guards Armoured* (Geoffrey Bles, 1956)

Saunders, Hilary, *Red Beret* (Michael Joseph, 1950)

Shulman, Milton, *Defeat in the West* (Secker & Warburg, 1947)

Speer, Albert, *Inside the Third Reich* (Sphere, 1971)

Stacey, C. P., *History of the Canadian Army* (Queens/Ottawa, 1960)

Taylor & Shaw, *Dictionary of the Third Reich* (Penguin, 1997)

Trevor-Roper, Hugh, *Hitler's Table Talk* (Enigma Books, NY 2000)

Whiting, Charles, *Monty's Greatest Gamble* (1985)

Wilmot, Chester, *Struggle for Europe* (Collins, 1957)

Also Divisional Histories: *Red Crown and Dragon, Polar Bears, Fighting Wessex Wyverns, The Black Bull, Churchill's Desert Rats, Monty's Ironsides, Monty's Northern Legions, Monty's Highlanders* by Patrick Delaforce.

Glossary

German and Dutch Terms

Abteilungen	Combat Units
Abwehr	Military intelligence department of the OKW
Festung	German Garrison designated only by Hitler
Flak Units	German anti-aircraft units
Hitler Jugend	Young and highly indoctrinated Nazis, fanatically loyal to Hitler
Kampfgruppe	Impromptu German battle group
Kranken	Wounded convalescent formation
Kriegsmarine	Members of the German Navy
Landsturm Nederland	Dutch Nazi volunteers
Luftwaffe	German Air Force
Minnenwerfer	German mortars known by the Allies as 'Moaning minnies'
Nebelwerfer	Mortar that fired chemical or HE shells
OKW	Oberkommando der Wehrmacht
Panzerfaust	Cheap German one-shot anti-tank weapon
Panzergrenadier	German motorised or mechanised infantry
Panzerjaeger	Armoured units designed to fight tanks, usually SP guns
Schmeisser	German submachine-gun, also known as the MP-40
Schu mine	Wooden cased anti-infantry mine
Schwerpunkt	Concentrated method of advance upon a focal point with co-ordinated armour and infantry and appropriate flank protection
Spandau	German machine-gun
Stonk	German word for a sudden artillery bombardment, also used by the British
Teller	Cylindrical, powerful anti-tank mine

Verdronken land	Flooded region in Holland
Volksgrenadier Divisions	Emergency divisions created in 1944 to combat the German manpower crisis
Wehrmacht	The unified armed forces of Germany

Military Abbreviations

AA	Anti-Aircraft
ADC	Aide-de-camp
AFV	Armoured fighting vehicle
AVRE	Armoured Vehicle Royal Engineers; special Churchill tank
BAOR	British Army of the Rhine
BEF	British Expeditionary Force
BLA	British Liberation Army
CAS	Close Air Support, usually by RAF planes
CIGS	Chief of the Imperial General Staff
CO	Commanding Officer
CRA	Commander Royal Artillery
CRE	Commander Royal Engineers
Crocodile	Churchill tank with flame-thrower
CSM	Company Sergeant-Major
DCLI	Duke of Cornwall's Light Infantry
DCM	Distinguished Conduct Medal
DD Tanks	Duplex Drive Tanks
DLI	Durham Light Infantry
DP	Displaced Person
DSO	Distinguished Service Order
DUKW	Wheeled amphibious landing craft
DZ	Drop zone for paratroops
FOO	Artillery Forward Observation Officer
GOC	General Officer Commanding
HAA	Heavy anti-aircraft gun
HLI	Highland Light Infantry
Kangaroo	Armoured infantry carrier on selfpropelled tank-type chassis
KOSB	King's Own Scottish Borderers
KOYLI	King's Own Yorkshire Light Infantry
KRRC	King's Royal Rifle Corps
KSLI	King's Shropshire Light Infantry
LAA	Light Anti-Aircraft artillery
LCA	Landing Craft Assault
LZ	Landing zone for glider troops
MC	Military Cross
MM	Military Medal

MMG	Medium Machine Guns
NAAFI	Navy, Army and Air Force Institutes
NCO	Non-Commissioned Officer
OBE	Order of the British Empire
OC	Officer commanding
PBI	Poor Bloody Infantry
PIAT	Projector Infantry Anti-Tank
PIR	Parachute Infantry Regiments
RA	Royal Artillery
RAC	Royal Armoured Corps
RAMC	Royal Army Medical Corps
RB	Rifle Brigade
RE	Royal Engineers
REME	Royal Electrical and Mechanical Engineers
RHA	Royal Horse Artillery
RSF	Royal Scots Fusiliers
RTR	Royal Tank Regiment
RWF	Royal Welch Fusiliers
SHAEF	Supreme Headquarters Allied Expeditionary Force
SOE	Special Operations Executive
TCV	Troop Carrying Vehicle
VE Day	Victory in Europe Day
UNRRA	United Nations Relief & Rehabilitation Administration

Index